D0161824

FEAR NO MORE

AN ADULT APPROACH TO MATHEMATICS

Peter Hilton
State University of New York at Binghamton, and
Batelle Memorial Institute, Seattle, Washington

Jean Pedersen
University of Santa Clara, Santa Clara, California

Addison-Wesley Publishing Company
Menlo Park, California   Reading, Massachusetts   London
Amsterdam   Don Mills, Ontario   Sydney

To our dear friend George Pólya

This book is published by the Addison-Wesley Innovative Division.

Copyright © 1983 by Addison-Wesley Publishing Company, Inc. All rights reserved. No part of this publication may be reproduced, stored in a retrieval system, or transmitted, in any form or by any means, electronic, mechanical, photocopying, recording, or otherwise, without the prior written permission of the publisher. Printed in the United States of America. Published simultaneously in Canada.

ISBN-0-201-05713-1
ABCDEFGHIJ-DO-898765432

Preface

MATH AVOIDANCE IS A SERIOUS malaise of our time. It has many causes which can be effectively grouped under three headings: societal, familial, and cultural influences; pedagogy; and curriculum.

This book was conceived out of the desire to find a cure for the malaise based on a new approach to pedagogy and curriculum. For it is our conviction that much of the problem lies in these domains. We may therefore leave to others the formidable task of eliminating the first of the causes listed above. This certainly does not mean that our task is easy, but it does mean that the changes we want to make are under our own control. We believe that this book responds to a very real need, and feel it does so in a way that makes it unique among the efforts to avoid math avoidance.

We go into greater detail about our pedagogical and curricular reforms in Chapter 0, which, in addition to serving as an introduction to the rest of the text, plays, in our view, an essential role in preparing the student to approach the problem of learning mathematics in a mature, adult way. We see the publication of this book as providing the opportunity to test the validity of the pedagogical principles we espouse and the good sense of the reforms we recommend.

It is a pleasure to acknowledge the help we have received from many sources. First, we are grateful to the University of Santa Clara for making it possible for us to conduct a pilot course based on a preliminary version of this text. We should mention in this regard the names of Fr. Paul Locatelli, S. J., Academic Vice-President; Dean John Drahmann; Professor Gerald Alexanderson, chairman of the Department of Mathematics; and Professor Mary Gordon, Director of the Women's Studies Program. Second, we are especially grateful to the very articulate students who volunteered to take that course and who did not hesitate to give us the benefit of their constructive criticism. We want to acknowledge the assistance and encouragement we have received from the Innovative Division of Addison-Wesley, which enabled us (indeed, obliged us!) to produce this text with the maximum of dispatch. We have received valuable advice and support from many people professionally concerned with the problem of math anxiety and from mathematics educators. We would particularly like to mention, in the first category, Sheila Tobias, in the second, Frank Armbruster, and — in both — Dena Patterson.

It is a particular pleasure to express our appreciation of the encouragement we have received from George Pólya, from the moment we conceived the idea of writing this book until the day when we were able to place the finished manuscript in his lap. We wish to thank the reviewers of our manuscript for making several very valuable suggestions which we were happy to incorporate into the final version. We are especially grateful to Dave Logothetti for his invaluable advice about illustrations and for his trenchant comments on our idiosyncratic punctuation.† We also wish to express our appreciation to another colleague, Gerald Alexanderson, for his suave and urbane comments which performed a function that bears his unmistakable stamp. Finally, we express our gratitude to our families who accepted, with understanding and patience, all the inconveniences and difficulties attendant on our preoccupation with the writing of this book.

† !

Contents

CHAPTER

0

Why Should You Study Mathematics— And How?

We strongly advise you to read the first 4 sections of this chapter (the first 5 if you are studying on your own) before embarking on the rest of the text. If you prefer, you could postpone the study of sections 0.6, 0.7, and 0.8 until you have covered the content of Chapters 1, 2, and 3.

0.1 MATHEMATICS IN CONTEMPORARY LIFE

Today, more than ever before, mathematics enters the lives of every one of us. Thirty years ago it was supposed that mathematics was only needed by somebody planning to work in one of the 'hard' sciences (especially physics and chemistry), or to become an engineer, a professional statistician, an actuary or an accountant. Today, however, there are very few professions in which an understanding of mathematics is irrelevant. In the biological sciences, in the social sciences (especially economics, town planning and psychology), in medicine, in business administration, mathematical methods of some sophistication are being used. Indeed, in the everyday life of today's citizen, there are numerous occasions on which some understanding of mathematics, and of the technology based on it, is extremely vital, and on which the lack of that understanding leaves the individual with a feeling of helplessness, incompetence and bewilderment. The computer and its baby brother, the hand calculator, invade our lives constantly, presenting us with statements, bills, meter readings, inventories; and many of us quail before this onslaught and sigh nostalgically for the good old days when you didn't need to understand decimals or approximations. But those days will not return!

The overall importance of mathematics in our lives is reflected in the curricula of our institutions of higher education. Some mathematics is now prerequisite for almost all major disciplines at the college level. Students who do not take three years of high-school mathematics find themselves very severely handicapped on entering college—if they are not to be excluded from a vast range of possible majors they have to undertake the dauntingly difficult task of making up their deficiencies in the course of their first two years of college work. Regrettably, many of today's students are in precisely this unfortunate and uncomfortable position; and this serious societal problem has the additional aspect of affecting women significantly more than men. One of the best researchers in this field of sociological enquiry, Lucy Sells, has drawn attention to the fact that, in one year in the 1970's, only 9% of the entering freshwomen at the University of California, Berkeley, had had twelfth grade mathematics, whereas the figure for freshmen was 52%. Our reaction to this is that the figure for men is bad enough; the figure for women is appalling.

For the true irony of the situation is that the increasing importance of mathematical understanding in our professional and daily lives, and the increasing awareness of that importance, is *not* being matched by an increase in mathematical understanding itself. On the contrary, statistical evidence (provided, for example, by the scores obtained by the National Assessment of Educational Progress from tests administered at regular intervals to samples of 9-year-olds, 13-year-olds and 17-year-olds) and subjective experience both lead to the unavoidable conclusion that the level of mathematical comprehension is actually declining in our population.† Moreover, various societal forces (for example, peer opinion, parental influence, the guidance of school counselors) act particularly to dissuade women from pursuing courses of study, and careers, that call for some mathematical competence and understanding. Thus the disadvantages from which women suffer in our society are being compounded, just at the time when we claim to be particularly zealous in our attempts to eliminate them.

Others far more expert than we have studied these sociological factors in some detail; we particularly commend the first four chapters of Sheila Tobias's book *Overcoming Math Anxiety* (Norton, 1978) in this regard, together with the excellent bibliography she provides. Our concern in this book (the first of a projected series of three volumes) is to provide what we regard as the appropriate mathematics for today's citizens. We stress that we have in mind not only appropriate mathematical content but also appropriate *style of*

---

†It appears that competence with the drill problems of elementary arithmetic may be at a fairly stable, if unimpressive, level.

presentation. For it is our claim that mathematics education fails today, and on a massive scale, for many reasons. Some have to do with societal forces (among them, for example, the gross misuse of the television medium by commercial interests, and the consequent disastrous effect on the entire learning process of our children); but, within mathematics education itself, there is much that is wrong with the choice of content of (so-called) mathematics courses and perhaps even more that is wrong with the way in which the material is presented to the student. However well-meaning the intention and devoted the effort, the attempt to reteach the same unattractive material, in the same repellent fashion, to those adults who rebelled against it as children is doomed to failure; and that failure leaves a permanent mark on all who have exposed themselves to the experience.

We propose, in this chapter, to describe in some detail our approach to the teaching of mathematics to intelligent adults. First, however, it is our responsibility to document our case that something new and different really is needed.

0.2 THE STANDARD MATHEMATICS CURRICULUM AT THE PRE-COLLEGE LEVEL

What is mathematics? We do not propose to offer a formal definition, but let us at any rate describe some of its characteristic features. We use mathematics in order to analyze systematically the processes going on around us, both natural processes and man-made processes. We set up a mathematical model of a given situation by abstracting from that situation, stripping the situation of those features that are not relevant to the question we are asking. Thus if we need to know the average daily rainfall in Springfield, Massachusetts, in February, 1980, we consult the records for the rainfall for each day of February. We then record the number of inches per square foot (if that is the unit we are using or the unit in which the rainfall was recorded) and we are left with the arithmetical task of adding together 29 numbers (some of which might be zero) and dividing the sum by 29 to get the average. We do not need to take into our mathematical model the fact that the measurements were in inches per square foot; we do not even need to know we are talking of rainfall and there is nothing in the mathematical model to tell us we are talking about Springfield, Massachusetts! Only when we get our arithmetical result must we remember what it means in the special context of our problem.

Thus, mathematics involves, in an essential way, *abstraction, analytical reasoning, computation* and *interpretation.* The actual computations themselves are, in a certain respect, the least *mathematical* part of the whole procedure; they are certainly the least *human* part, since machines can carry out computations far more effectively than human beings. What is essentially human is deciding what computation to do, and mathematics has developed in response to our awareness of the overriding importance of making it the best possible analytical tool for thinking about the real world. Mathematical concepts fall effectively into two categories, reflecting our way of organizing our experiences; the first is the category of *number* (arithmetic and algebra), the second is the category of *shape* (geometry). These two fundamental categories are brought together through the unifying concept of *measurement,* which enables us to express key properties of shape (length, area, volume, angle, etc.) in terms of an extended notion of number (through *decimals* or *fractions*). Since we are not content to make a *static* study of our world, but wish to understand it as a *process,* we must refine our mathematics to abstract from that world of process the concept of *change* and this leads us to the *calculus,* which is nothing but the study of continuous change. Since our knowledge of the real world is almost always incomplete, we often have to make intelligent guesses, so that we need a corresponding mathematical theory, and this is *probability theory,* again using the extended concept of number.

How absurd it is, then, that the traditional elementary syllabus puts such huge emphasis on computation, often virtually ignoring the most characteristic, *thinking* parts of mathematics. Children are asked to learn various dull arithmetical algorithms, but are seldom carefully instructed in *why* these algorithms are important, *why* they work, *why* they were chosen and *how* they reflect natural processes. Moreover, the 'answer' is usually just a meaningless number, whereas the 'real' answer is usually a new, or enriched, insight into the real world, on the basis of which a rational decision can be made. Very often such decisions only require an *approximate* solution to the arithmetical problem—there is no *intrinsic* advantage in a precise calculation over a careful estimate and the latter may be both easier and more useful.

Yet we find a dreadful preoccupation with drill calculations pervading the standard curriculum. There is even a powerful movement in mathematical education today, called the 'Back to Basics' movement, which advocates this preoccupation. Its supporters draw attention to the (undoubted) fact that many children don't even know their basic number facts, and infer from this that recent pedagogy has been at fault in laying too much stress on understanding at the expense of basic skills. We contend (a) that ignorance of basic number facts is not due to the prevalence of the 'new math', which never reached more than 15% of our children—rather it is due to the prevailing attitude to education and to learning; and (b) that what the 'Back to Basics' movement identifies as basic skills—that is, the traditional arithmetical algorithms with whole numbers—never were basic and are even less basic today than they ever were! We must go *forward,* not *back,* if we are to fashion an education worthy of our society and one which will enable our citizens to go forth confidently into our changing world able and willing to play their part and to meet its challenges.

We contend that the teaching styles of the standard mathematics curricula, and the curricula recommended by the 'Back to Basics' brigade in particular, are guilty of two fundamental (and related) faults, which we describe as *authoritarianism* and *mystification.* They are authoritarian in that the student is simply told what to do without adequate explanation—perhaps even without any explanation. The arithmetical problem usually has no source and the method of solution is given as a rule to be remembered. Soon the student's memory becomes overloaded, the technique is misremembered or totally forgotten and, in the absence of understanding, cannot be reconstructed. The result is incompetence, discomfort, distaste. The teaching often mystifies where it should enlighten. The student is led to believe that he or she should not even *try* to understand—that way lies confusion. Perhaps later enlightenment will come; for now, the student should accept the 'fact' that mathematics is a mysterious process and simply learn to use it skillfully. Moreover processes are introduced for which the student has no real use so that mathematics looks like a procedure for obtaining meaningless answers to meaningless questions by meaningless calculations.

It would seem obvious that this is no way to teach mathematics—or anything else for that matter. Yet it is our experience that this *is* the way mathematics is often taught—and even at the college level. Let us give one example, drawn from our own experience, and relating to the work described in Chapter 5 on negative numbers. We are concerned there with the multiplication of negative numbers and the constantly recurring question—why should the product of two negative numbers be positive? We offer in Chapter 5 a careful explanation of why it is *useful* to define the product of two negative numbers to be positive, insisting at the same time that, for most people, it wouldn't ever be necessary to multiply together two negative numbers. When we presented this approach at a recent workshop for mathematics teachers in two-year colleges, we were met (on the part of some of the participants) with the reaction that (a) they wouldn't have time for so long and involved an explanation, (b) many students wouldn't understand it anyway, (c) they had to teach the product of two negative numbers because it was on the syllabus, and (d) what was wrong anyway with simply *telling* the students that the product of two negatives is positive?

In this example, a problem rears its head which lies at the heart of the question of how to improve the teaching of mathematics—the tyranny of the 'syllabus'. A certain amount of material must be covered in the allotted time, that material is prescribed in advance;

and that material is going to be tested. We declare categorically that no real improvement can be achieved so long as we continue to test our students as we do today. 'Competency-based criteria' have nothing in common with mathematics, however relevant they may be to computational skill. The attempt to analyze progress in mathematics into a set of skills which can be described as the 'objectives' of a given course is misconceived and wrong-headed. Standardized tests distort the curriculum, emphasize mechanical skills at the expense of understanding, and impose on the curriculum and teaching style an inertia to change so damaging as to be, in many cases, fatal. Crucial decisions about the teaching of mathematics are not to be put into the hands of bureaucrats and so-called 'education experts'— unless the latter are genuinely expert in mathematics and mathematics education. The syllabus should always be kept as flexible as possible, to meet the needs of the students.

This book is designed for adults, not children. This releases us from the necessity to prepare the reader for the kinds of tests our unfortunate children have to submit to, and from the necessity of covering a prescribed syllabus. We do maintain, however, that our readers will have the chance, through a study of our text, of achieving a mastery of the fundamentals of arithmetic and thus even of achieving success if faced with the kind of test that we regard as irrelevant to the acquisition of real understanding of the role of mathematics. We now describe our own approach as reflected in this book.

0.3 THE APPROACH TAKEN IN THIS BOOK

Mathematics is immensely useful—otherwise it would not exist. It follows that it must be taught to emphasize its usefulness. This does not imply that careful attention should not also be given to its internal structure; for that internal structure gives the mathematics its power and thus enhances its utility. Moreover, an absolutely essential feature of mathematics is its *universality*; the same piece of mathematics may be the appropriate tool in widely different real-world contexts. This principle is familiar with such elementary concepts as the addition and multiplication of whole numbers; but it also holds, of course, in more advanced areas of mathematics. For example the differential calculus (which will be discussed in the third volume of the series) enables us to study any situation of continuous change, or, indeed, any situation which can be reasonably approximated by one of continuous change. Thus this powerful mathematical tool may be applied to study the path of a bullet fired from a pistol, the expected inflation rate in the year 2000, or the trend in the fish population off the California coast. Again certain parts of linear algebra (to which we introduce the reader in the second volume) may help the traveling salesman to plan his itinerary, the television set manufacturer to plan a sequence of modifications of her standard models, or the cryptographer to decipher the coded message of the enemy.

What follows from these basic aspects of mathematics? The utility aspect implies that mathematical concepts and operations should be presented as entirely *natural,* responding to our desire to study our world systematically and to derive information about it readily and as easily as possible. We do not study decimals because 'we've done fractions'; we study decimals because we use a decimal coinage system and, increasingly, a decimal (metric) system of weights and measures, and because that essential tool of modern living, the calculator, uses decimals. We do not study 'geometry' because 'it's in the syllabus'; we study geometry because the basic concepts of geometry reflect essential features of the three-dimensional world in which we live, and because we can best represent changing phenomena pictorially as graphs.

But the universality of mathematics dictates that we should not teach a piece of mathematics in too close connection to a particular application, since this may leave the student with the impression that this is its sole justification. The student will then find it very difficult to *abstract* the mathematics from the concrete application and subsequently apply it to a totally different problem. Thus the mathematics, once brought into existence to solve a particular problem, should then be studied *for itself,* so that the techniques developed are then available in other contexts.

Let us give one very fundamental example of this point. It is perfectly acceptable, of course, initially to present the addition of two whole numbers as the mathematical model appropriate to the counting of objects in two disjoint collections. Thus if we have 7 balls in one bag and 12 balls in another bag, then we can determine how many balls we have altogether by performing the sum $7 + 12$ (and obtaining the answer 19). However, if we *only* think of the addition this way we are seriously handicapped. For it is also true that we should use addition if presented with the problem that we start 7 miles south of our camp and walk 12 miles in a southerly direction and wish to know how far south of our camp we are as a result. Here addition is 'modeling' a very different situation. Moreover, how are we to extend our rules of addition to decimals, or fractions, if we only think of counting the number of objects in a disjoint union? If one rock weighs 12.6 pounds and another 13.1 pounds, the total weight is 25.7 pounds obtained from adding the decimals. But we cannot meaningfully talk of taking first 12.6 objects and then 13.1 objects and counting how many objects we have altogether!

The 'New Math' put great emphasis on explaining the meaning of elementary mathematics, and this principle is surely to be commended. However *in practice,* the 'New Math' often went wrong in precisely the way we have described above. A *single* interpretation (via set-theoretical union) was used to explain the meaning of addition and this locked the student (and often, the teacher) into a highly restrictive approach to addition, which was ultimately extremely unfortunate.

The emphasis we give to particular parts of mathematics in this book is based on their importance and not on their traditional role in the curriculum. Thus we guide the student through the material, explaining both the mathematical ideas themselves and their significance in applications, and we do not hesitate to admit when a particular topic should be regarded as minor. Two such topics deserve special mention here, since they are topics usually given very great attention.

The first is the *addition of fractions.* As an arithmetical operation, the addition of fractions (unlike their *multiplication*) is awkward and clumsy. Moreover it is an operation that is *rarely applied.* Broadly one can say that one adds decimals, not fractions, and the addition of decimals is easy and natural. But, for hundreds of years, long-suffering students have been expected to learn how to add fractions. This has been taught almost exclusively as a 'skill', divorced from all applications. It has therefore been learned as an effort of memory, and, almost inevitably, later forgotten. It continues to appear on standard tests, and thus continues to enjoy a totally undeserved prominence in the curriculum. Certainly we explain how to add fractions, but our approach is low key and we do not 'oversell the product'. We motivate the addition of fractions largely through elementary ideas of probability theory, where it appears as a fairly natural operation, but by no means in its full—and unappealing—complexity.

The second topic which we treat very mildly is that of the multiplication of negative numbers, to which we have already referred in this chapter, and about which we need say little now. Both these topics come into their own when we come to study algebra in its full force (in the second volume) and, in our view, that is the time to emphasize them. We do not believe in the 'promised land' style of pedagogy whereby the teacher, or the text, promises the student untold and unimaginable delights later if he or she will only agree to suffer the discomforts of meaninglessness and incomprehension in the present.

Thus the emphasis we give to the topics we treat in this volume is dictated by their importance to the reader as an intelligent citizen of today's world, taking into account the stage of mathematical education he or she has reached. We are not constrained by the demands of some arbitrary and archaic examining syllabus in choosing our material. Nevertheless, we are forced, of course, to take into account the fact that our readers will doubtless be obliged to submit to standard tests of their mathematical skill, and that these tests will feature prominently the horrendous algorithms which remain the bane of our long-suffering children. We take account of this needless imperfection in our reader's world by providing an appendix describing in detail the standard algorithms of the arithmetic of whole numbers. We also give our readers plenty of opportunity to acquire skill in the manipulation of decimals, fractions and negative numbers, by providing exercises at the end of appropriate sections of the chapters dealing with these topics. But we believe that *skill and understanding go hand-in-hand.* If skill is taught without understanding the skill is quickly lost and forgotten; it may be retained just long enough for the student to pass some test, but it will not be *useful* to the student as it will not be available when it is actually needed. But if the skill grows out of the understanding (aided, of course, by the necessary practice) then, even if it is temporarily lost, it can be recovered. Thus we have a *practical* argument for recommending that we should teach for understanding and not merely for mechanical skill, to set alongside the *humanistic* argument that we are educating intelligent human beings and not training robots. That we are fully alive to the practical realities of the difficulties facing today's students in not only acquiring mathematical proficiency but also in passing tests of dubious relevance, often designed by those whose expertise does not lie in mathematics, in mathematical applications or in mathematical education—that we are alive to these realities should, we believe, become plain from the list of topics covered in this volume.

0.4   OUTLINE OF CHAPTER CONTENTS

Following this introductory chapter, which we regard as an integral part of the book, there are two very basic chapters, the first on *Decimals,* the second on *Fractions.* The order is important here—and non-traditional. For decimals arise as a perfectly natural extension of our number system in its most fundamental usages, that is, for counting and measuring. The simplicity of the arithmetic of decimals, compared with the arithmetic of fractions, derives from the nature of this extension. Unfortunately, the arithmetic of decimals does not yet play its full role in our civilization because our society has not yet agreed to use the metric system for all measures—we use a decimal coinage, but we continue to measure distances in the awkward units of inches, feet, yards and miles, and weight in the equally awkward units of ounces, pounds, hundredweights and tons.

We explain in Sections 1 and 2 of Chapter 1 how our decimal coinage and the metric (decimal) system of weights and measures lead naturally to decimal numbers, and how easy much of the arithmetic of decimals is—indeed, most of decimal arithmetic is really nothing but the arithmetic of whole numbers, as we explain. The multiplication of decimals does introduce a new idea into arithmetic, so we devote a section to discussing it. The final section of Chapter 1 is concerned with the theory and practice of *percentages,* a very important concept in our use of arithmetic to describe many features of the world around us, from population trends to sales taxes. It is appropriate to take up this idea in this chapter, because, from the *mathematical* point of view, percentages are just decimals 'thinly disguised'. We do *not* discuss division by decimals in this chapter, because division (by whole numbers, decimals or fractions) presents special difficulties and so merits a chapter all to itself (Chapter 3). Similarly, we say little about *approximate* calculations involving decimals in Chapter 1, since this important topic is taken up in Chapter 4.

Chapter 2 is, as we have said, concerned with fractions. The arithmetic of fractions is substantially more difficult than the arithmetic of decimals—this alone would justify our treating decimals first, but we have also given other reasons in the previous discussion. Decimals constitute an *immediate* extension of the number concept as applied to counting and measuring; fractions, on the other hand, do not first make their appearance in arithmetic as numbers, but as descriptions of *parts or proportions,* 'half a loaf', 'three quarters of an hour'. Thus our first understanding of fractions must be in this sense. The first arithmetic *operation* we meet involving fractions deals with fractions of a whole number—thus, $\frac{2}{3}$ of 12 is 8, and we should know how to obtain such fractions of numbers. Next we meet the idea of 'fractions of fractions'; a half of a quarter is an eighth and, from this key idea of fractions of fractions, we are led naturally to the idea of *multiplying* fractions.

It may happen that two *different* fractions lead to the *same* amounts; thus $\frac{2}{6}$ of 60 is the same as $\frac{1}{3}$ of 60, and $\frac{2}{6}$ of a meter stick has the same length as $\frac{1}{3}$ of a meter stick. When different fractions produce the same amounts they are called equivalent. So $\frac{2}{6}$ is equivalent to $\frac{1}{3}$, $\frac{3}{30}$ is equivalent to $\frac{1}{10}$, $\frac{9}{300}$ is equivalent to $\frac{3}{100}$.

These notions constitute the *elementary* theory of fractions. The theory becomes far subtler, and more powerful, when fractions themselves *become numbers.*† This transition may be effected by thinking of numbers in their role as measurements. Thus $\frac{2}{3}$ of a yard is 2 feet, so, if we have a stick 2 feet long, we can say that, measuring the stick in yards, the length of the stick is $\frac{2}{3}$ yds, or *the number of*

---

† To speak very precisely—as we sometimes should, but not always!—fractions represent numbers, equivalent fractions representing the same number.

yards in the stick is $\frac{2}{3}$. Similarly, for a stick 5 feet long, the number of yards is $\frac{5}{3}$. Here we find use for fractions bigger than 1–a most important extension. We also find that we can regard $\frac{5}{3}$ as the result of dividing 5 by 3, itself a basic extension of the division concept.

Once fractions can be regarded as numbers, it becomes meaningful, and valuable, to compare fractions and decimals, and this is the topic of Section 4. Of course, our extension of the number concept to fractions does *not* imply that fractions have lost their original meaning as parts–they have taken on a new, enlarged, enriched meaning which does not invalidate their original significance. We are still free to regard a cake, or an idea, as half-baked without having to think of 'a half' as a number like 2 or 3! This remark is important, since it is a common error to suppose that, with increased mathematical sophistication, early mathematical ideas lose their validity. The student who says that parallel lines never meet should not feel out-faced by the older student who announces that parallel lines meet at infinity–both statements are true and only the context can determine which is the more useful formulation. This general point is further brought out in Chapter 4; it is clearly necessary to stress it here, since extremely intelligent non-specialists, even of the caliber of Sheila Tobias, go wrong on just this point, believing that the broadening of the scope of a mathematical concept introduces ambiguity, whereas such generalization is always mathematically consistent.

It is not absolutely accurate to say that fractions can be regarded as numbers. We have already discussed *equivalent* fractions; since equivalent fractions yield the same *amount,* they represent the same number. Thus the fractions $\frac{4}{6}$ and $\frac{2}{3}$, which are *different as fractions,* represent the *same number.* Such a number is called a *rational number,* to distinguish it from those numbers (like $\sqrt{2}$, π) which cannot be represented by fractions (though *all* numbers can be approximated by fractions). So we may think of the individual fractions of a set of equivalent fractions as being different *names* of the same number, just as 'Shah', 'Reza Pahlavi' and 'last occupant of the Peacock Throne' are all names of the same person, the recent absolute ruler of Iran.

Section 5 deals with the dread subject of the addition of fractions. As we have said, this is a relatively unimportant aspect of the arithmetic of fractions, elevated to an undeserved prominence in the traditional curriculum. We motivate the addition of fractions as an aspect of *elementary probability theory*; this application, however, in no way serves to justify the extraordinary, and clumsy, arithmetical contortions associated with such artificial problems as $\frac{7}{13} + \frac{12}{47}$. For us a typical problem is $\frac{2}{36} + \frac{3}{36}$; notice that we do not present the problem as one involving *reduced* fractions ($\frac{1}{18} + \frac{1}{12}$), since, in the applications to probability theory, it typically occurs that the fractions involved in the sum are naturally expressed with the same denominator.

Our discussion of probability inevitably leads us to introduce, in Section 7, the idea of odds. We pursue this topic rather far, since it is basic to the scientific method. Thus this section is the first in the volume to be adorned with an asterisk (*), indicating that the material it contains is at a higher level of difficulty. The reader may choose to ignore such a starred section on first reading; alternatively, the reader may be content to peruse the content of such a section more lightly, less intensely, than that of an (allegedly) easier section, being satisfied to absorb its flavor, on first reading, rather than to study it for mastery.

The final section of Chapter 2 discusses the notion of *average* as an application of fractions. Here the conceptual side tends to be stressed, as we point out the dangers of crude arguments involving averages and stress the usually neglected distinction between *fractions* and *rational numbers*.

Chapter 3 deals with the arithmetical operation of *Division*. We will not say more of the content of the first two sections of this chapter beyond remarking that they draw attention to the real difficulties of getting meaningful, and correct, answers to real-world problems involving division—difficulties not related to the problem of employing with accuracy the horrendous algorithm usually inflicted on our long-suffering children—and also to the silly stuff usually to be found in elementary texts when discussing division problems. We claim that the traditional mishandling of division is a basic cause of math anxiety and a perfectly adequate justification for mass avoidance of (so-called) mathematics.

The next two sections of this chapter deal with the rules for dividing by decimals and by fractions, and with applications of these operations. Again, our touch is light, since these processes occur far more rarely than those of *multiplication* by decimals and by fractions, topics to which we devote appropriately substantial attention. The final section summarizes and justifies certain arithmetical identities which will have appeared empirically in our treatment so far. For example, we explain why the identity

$$(A \times B) \div C = A \times (B \div C)$$

holds (whereas, of course, the identity $(A \times B) + C = A \times (B + C)$ is *almost always false*). This section is starred because it will appeal only to those who are interested in the mathematics for its own sake (but note that this is a larger group of people than those who are *only* interested in the mathematics for its own sake!).

Chapter 4 is concerned with *Estimation and Approximation.* These topics have many aspects. Within arithmetic itself, we often need to perform an approximate calculation to check the accuracy of a precise calculation. The advent of the handheld calculator has increased the importance of approximation, just as it has reduced– almost to vanishing point–the importance of the traditional algorithms of accurate calculation (we refer here to 'long' multiplication and division). For if an accurate calculation is required it can be performed on the hand-calculator painlessly; but it is prudent first to estimate the order of magnitude of the answer, so that we may check for such errors as punching the wrong key or misreading the problem when transferring it to the calculator.

However, it is important to realize that many problems are *solved completely* by means of an approximate calculation–that is, no precise calculation is required. 'I wish to buy 12 pens at 97¢ each, I have $10. Is that enough?' 'Can I afford a subscription of $7 a week if I am given $400 per year?' These and similar questions are answered by means of efficient arithmetical techniques which do not proceed via any accurate calculation. And if we learn to approximate and to estimate as a mathematical procedure, we must learn *when* to approximate and what *degree* of approximation is appropriate to a particular problem.

There is, moreover, a new arithmetic involved in taking approximate measurements. If the odometer on my bicycle shows distances to within tenths of a mile, and if it appears that I have cycled 7.2 miles from my home to my friend's house, the odometer having advanced from 187.3 to 194.5, what are, in fact, the limits within which the true distance must lie? If I cycle back by the identical route, will the odometer show 102.7 on my return? If not, what might it show? These questions, and questions like them, are carefully discussed in Chapter 4.

A further topic which falls naturally into this realm of ideas is that of *orders of magnitude,* and *scientific notation.* This last topic is very important, as the use of scientific notation (or, at least, of *exponential* notation) is essential for the presentation, and retention, of the significant parts of very large and very small numbers. It thus facilitates the comparison–that is, the approximate comparison–of such numbers; and we may talk of such numbers differing, if indeed they do, by orders of magnitude.

The subject of Chapter 5 is the nature and the arithmetic of *Negative Numbers.* Since negative numbers are used extensively in measurement, we believe that all intelligent adults should try to understand their meaning. So far as meaning is concerned, there is no extra difficulty, once negative integers (like -5, -7, -41) are understood, in extending the idea to negative decimals and negative fractions. We also recommend that the *addition* and *subtraction* of negative numbers be studied by everybody; however the *multiplication* and *division* of negative numbers will probably only be needed by those concerned with the physical sciences and engineering. Thus the most important section of Chapter 5 for the general reader is undoubtedly Section 5.1. However, there is also much of interest to the general reader in Section 5.2, since it presents an example of the progress of mathematical thought and the development of helpful, simplifying notation, at a comparatively elementary level.

Indeed this chapter exemplifies with unusual clarity our pedagogical principles and our sincere devotion to them. For, whereas the authoritarian approach simply instructs the student to 'learn' that 'a negative times a negative is positive', we devote many pages, and make a substantial digression into negative exponents, in order to show (a) that there may be situations in which it would seem reasonable to multiply by a negative number, and (b) that if we do want to multiply by a negative number, then the rule 'negative times negative is positive' turns out to be consistent with all our previous conventions.

Chapter 6, on *Factorization,* is largely included because we feel we owe it to the interested student to present the arithmetic of whole numbers, not merely as a tool in living a reasonable and intelligent life, but also as a part of mathematics. Again, the reader concerned exclusively with arithmetic as a tool may ignore this entire chapter. However, such a reader is warned at the same time that, if his or her taste does not extend to the material of this chapter (we speak here of *enjoying* the content, not *mastering* it), then it would not be advisable to pursue mathematics further than the content of this volume.

Since we are really *doing mathematics* in this chapter the style changes and becomes more precise, less informal. We announce results as *theorems,* and we *prove* them. However, the results we choose to discuss continue to be those accessible to the student we have had in mind all along—there is no jump here to totally new topics or totally new ideas. The motivations for the ideas presented—prime number, greatest common divisor, least common multiple, casting out 9's—are all to be found, and are explicitly described, in previous parts of this volume.

We are particularly pleased to have had the opportunity to present the topic of 'casting out 9's'. This is one of the most beautiful ideas of arithmetic. It is genuinely mathematical, but it is also of practical value. It enables one to perform a check on a complicated calculation which has the following important features:

(i)   if the calculation fails the test it is certainly wrong;

(ii) if the calculation passes the test it is an 'odds-on' chance that the calculation is correct. (More precisely, the odds on the calculation being correct are 9 times what they were before the check was carried out.)

Moreover, this type of check is the first example one meets of one of the most important ideas in the whole of mathematics—that of replacing a problem-situation by a simplified model whose 'inner structure' closely resembles that of the original situation. This contrasts sharply with the usual sense in which the term 'check' is used in elementary mathematics; the check is often just a different way of carrying out the required calculation—or, even worse, a comparison with the answer given in the back of the book.

The final chapter of this volume is an appendix on the *Whole Number Algorithms.* This is provided for the benefit of those readers who may feel they need to refresh their skills before embarking on the main body of the text, or who might wish to refer to those algorithms to strengthen their understanding of some particular part of the text. We do not view the material in this appendix as forming part of the course itself, so the treatment we give it here is very different from that accorded to the items in Chapters 1–6. We do, however, provide examples in the Appendix which can be used by the reader to establish his or her readiness for the main text material.

0.5 USING THIS TEXT FOR SELF-INSTRUCTION

We envisage that most readers of this text will be students enrolled in courses at community colleges, four-year colleges or universities. However, we have been very conscious, in our writing, of the obligation to serve the needs of those readers who are unable, for one reason or another, to enroll in such courses under the guidance of an instructor, and who are trying to raise their level of mathematical comprehension by their own unaided efforts. We address the reader directly in the text, and we try to anticipate the type of question that the reader is likely to ask. The style is informal, so that the reader may learn mathematics in an appropriately relaxed atmosphere. We try to signal to the reader what parts of the book are the most important and what parts are the most difficult. In this way we hope that the reader will not fall into the error of supposing that a failure to understand a mathematical idea *immediately,* or a failure to execute a complicated calculation with total accuracy, is a sign of ineffable stupidity! Be assured, dear reader, that the authors themselves have often found themselves failing in just these ways!

We also have tried to meet the needs of those who are trying to teach themselves by providing a large number of exercises, thus enabling the reader to monitor his or her own progress in skill and understanding. Answers are provided to these exercises, of course. We encourage the reader to attempt these exercises conscientiously and fairly systematically—mathematics is something to be *done* (not merely *learned*), and the proof that the idea has been grasped is that the reader is able to use it. There are special features of the exercises, described explicitly in the next section of this chapter, which we hope will make them more attractive and will enhance the value of attempting them systematically and comprehensively.

Our individual readers, teaching themselves by reading our book have, of course, the freedom to choose those parts they think they need. We are certainly not going to insist that this book be read from cover to cover! However, we do insist that it is neither a work of reference nor a compendium of results—it is a teaching instrument. Thus we do recommend that, if the material from any chapter is required by the reader, the whole chapter be read. This should certainly facilitate understanding—at the very least to the extent that notation and terminology will constitute no obstacle to the reader.

We have considered here the case of the individual reader, not enrolled in some regular mathematics course. But we have to say that we very much hope there will be no reader obliged not merely to be individual but also to be solitary. For mathematics is not best done alone. Mathematics is, among its many facets, a means of communication, in a sense a language. It is much better to test one's understanding by discussion with friends, to try to teach one's friends what one is in the process of learning. Mathematics should be done collaboratively, and so it should be learned the same way. If you are not enrolled in a course yourself, try to organize your own 'mini-course'—share your ideas, your thoughts, your attempts to understand and to solve problems.

Finally, we have to mention that there are, today, situations in which one is effectively in the position of receiving no regular instruction even though one is enrolled in a course! The shortage of mathematically trained teachers means that sometimes students find themselves in the situation of being 'taught mathematics' by unqualified people—football coaches or, perhaps, college gardeners. In such a situation the adult student has be behave with maturity—this means being decent to the instructor and letting the words of our text speak for themselves. In the final analysis, trust your judgment!

0.6 SPECIAL FEATURES OF THE TEXT

(a) The exercises The exercises are an integral part of our text. There comes a point when one is trying to explain an idea when it is useless to belabor it any further and one must say, 'Try it yourself to see if you've grasped it.'. The exercises are intended to fulfil this role—to provide the reader with the conviction that the idea has been grasped—or, as we must allow, the evidence that it has not. However, the exercises have also been designed to provide some fun for the student, and to develop the reader's capacity to recognize number patterns. Thus several sets of exercises have been devised in such a way that the answers to the exercises within a set reveal a pattern which the reader is asked to notice. Subsequent exercises may be based on that pattern. The patterns need not be related to the real-life or mathematical context of the problems in the set, though they often will be. We hope in this way, as we have said, to introduce a fun element even into those parts of our subject that are relatively computational and dry. But we hope to accomplish more; we hope that, by becoming skillful with the pattern puzzles which we present, the reader will become much more comfortable with numbers.

(b) Notation and terminology   What we are offering you is genuine mathematics, written in mathematical language with mathematical notation. It is our belief that much of the difficulty students encounter with mathematics is due to the use of bad, cumbersome notation. People tend to think that something 'new' happens with the advent of algebra, because, suddenly, one is using letters instead of numbers. All that is happening is that one is improving the language! Consider, for example, the following statement: 'If you take two numbers, add them, and double your answer, you get the same as you would if you doubled each number first and then added the results'. We contend that this is far harder to understand than the statement

$$2 \times (X + Y) = (2 \times X) + (2 \times Y).$$

Moreover this statement leads immediately to the natural generalization

$$n \times (X + Y) = (n \times X) + (n \times Y),$$

which is suggested by the very *form* of the first statement; but the words necessary to express the second formulation, without the use of algebraic notation, would constitute a truly horrendous sentence (try it!). Of course, you have to get used to statements like these, just as you have to learn any new language—or any set of instructions for a recently purchased household gadget. We believe this learning should take place as early as possible. Then there is no difficulty later on when the idea of a variable (that is, a symbol standing for any number from a given set, say the whole numbers, the integers, the fractions) may be used in a technically more advanced mathematical context.

When a number of variables appear in a natural sequence, it is reasonable to index them by the numbers 1, 2, 3, $\cdots$ (just as we do the rooms on a floor of a hotel). So if we want to consider ten arbitrary numbers we may use the notation $n_1, n_2, n_3, \cdots, n_{10}$

(where the dots are used simply to save effort, since it is perfectly clear exactly which six entries should go between n_3 and n_{10})— sometimes this is referred to as the notation of *subscripted variables*. In this case the numbers written below the normal line of print are the subscripts, and the n's, along with their subscripts, are the individual variables. The one warning to give the reader is that, whereas subscripting is just a method of indexing variables, super-scripts (numbers immediately following a variable, but written above the normal line of print) are exponents. Thus n^2 means $n \times n$, n^3 means $n \times n \times n$, etc. We cannot use the 'top right hand corner' for indexing—it has already been reserved for exponential notation.

In one respect we have tended in the text not to adopt standard mathematical notation. If the number n is multiplied by the number m the mathematician will write the result as mn: we have, almost always, written $m \times n$. Our reason is that there is little inconvenience in the expanded notation and it makes the transition to specified numbers easier. Thus if $m = 7$ and $n = 6$, it is obvious that $m \times n = 7 \times 6 = 42$. But there is a danger of writing 76 for the result of substituting $m = 7$, $n = 6$ into mn. Of course, we dare not omit the '$\times$' symbol when specific numbers are involved (we should not, either, when the numbers are decimals or fractions). Since our concern in this volume is primarily but not exclusively with arithmetic and we deal an enormous amount with specific numbers, we have not felt it worthwhile to impose the more sophisticated notation on our readers. But those who wish to adopt it may, of course, do so—this would be, in any case, a good introduction to the subsequent volumes.

There is a particular feature about our terminology to which we wish to draw attention, since it reflects an important pedagogical principle. Elementary texts use the phrases 'whole numbers', 'counting numbers' and 'natural numbers'. The question at issue seems to be the status of 0, zero. It is undoubtedly a whole number. Many claim that it is not a counting number and opinion seems to be divided as to whether it is a natural number!† We use the phrase 'whole number' in this text so that 0 is definitely included; and we recommend that 0 should be included in the set of natural numbers when they are used, for example, as the spring-board for the development of further extensions of the number concept. Notice that we use 0 as the number of *this* introductory chapter—this seems, to us, very accurately to reflect its (rightful) place in our text.

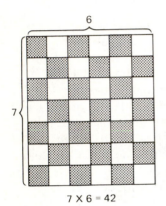

6

7

7 X 6 = 42

---

†This divided opinion is splendidly exemplified in Griffiths and Hilton, *Classical Mathematics,* Springer (1979), where on page 4 it is explicitly omitted from the set of natural numbers and on page 373 just as explicitly included!

It is no mere terminological convention that 0 is included among the whole numbers. The addition and multiplication facts should include statements like $0 + 7 = 7$ or $0 \times 5 = 0$, and these should, of course, be available for instant recall. By the same token it is *not* necessary to include the '10 times table' in the multiplication facts. For the rule of multiplying by 10 (namely, put a zero on the right of the number being multiplied) is part of the science of arithmetic and is not a 'fact' to be remembered like $9 \times 7 = 63$.

(c) Emphasis on labor-saving Even our remarks on notation carry the message that we hope has been conveyed in many parts of this chapter and, even more, in the text itself—good mathematics is labor-saving. Thus good notation is designed to avoid unnecessary thought—the notation 'takes charge' and dictates the appropriate arithmetical procedures. Thus, for example, we have an easy algorithm for adding 49 and 97; try adding these numbers using Roman numerals: XLIX + XCVII!

This feature of mathematics is rarely brought out in any text. Indeed, the opposite impression is often given—that mathematics creates work, that mathematical problems are something extra, and burdensome, that we have to do, and that we wouldn't have to do if there were no mathematics. Instead, we emphasize that mathematics is an aid which we can and should use to solve real-world problems and to understand real-world situations. If you want to know the area of a rectangular field in square yards, you can, of course, stake it out with posts at yard intervals the length and breadth of the field (not just along the perimeter) and then count the number of squares. It is easier just to measure the length and breadth in yards and multiply to get the area in square yards. If you want to know whether you can afford to eat regularly once a week in your favorite restaurant, you can simply go on eating there till you're heavily in debt. An easier and less drastic procedure, if perhaps not so pleasant gastronomically, is to estimate the cost of a meal and assess how much you could set aside from your net earnings for eating out.

Thus it is our purpose to make mathematics your friend, so that you recognize that it is helpful to you in your lives and, whatever your earlier experiences with mathematical ideas, well within your scope. Moreover, *we* invent mathematics so that it should be useful, and we invent mathematical notation to make calculation easier and ideas more perspicuous.

(d) Examples and general statements Bearing in mind the type of reader we are catering to, we do depart in one important respect from customary mathematical exposition. Suppose we wish to show why a particular result is true. We would first state the result in a particular case, for example,

$$\frac{3}{5} \div \frac{2}{3} = \frac{3}{5} \times \frac{3}{2} .$$

Here the 'result' is that to divide by a fraction is equivalent to multiplying by that fraction 'turned upside down'. We would explain why this example holds and would then remark that there was nothing special here about our choice of fractions so that our explanation would have applied equally well to any other fractions. Only then would we make the general statement

$$\frac{a}{b} \div \frac{c}{d} = \frac{a}{b} \times \frac{d}{c},$$

and we would not prove this last statement explicitly. This approach is more informal than a standard textbook for mathematics students, but we believe it will commend itself to our readers.

We cannot compare this approach with a standard textbook *at this level,* since such texts do not contain proofs in any sense. We do believe in presenting our readers with mathematical ideas, expressed mathematically. But, as with any language, it takes time to become familiar with its use, and to be able effortlessly to use it properly; we should not presume thorough familiarity with mathematical notation, nor ease with the handling of mathematical generality at this stage. We are not concerned to satisfy the pedants and the formalists but to make the best approach to our readers.

(e) Starred passages marked with a wavy rule   Readers of this volume will have different backgrounds, different motives, different abilities, different aspirations. Some will not be taking their mathematics beyond the level of this text; some will, we hope, decide at a relatively late stage of their study of this material that they would like to go further; some, the really ambitious ones, will have planned to go further even before beginning this volume; and some will be merely using this volume for private reading as a refresher for their further studies.

It follows that different readers will be suited to different paces, and to different choices of topic. It would be foolish for us to suppose we can supply explicit means for the fine judgments each individual reader must make for himself or herself. But we have incorporated one device which we hope will be helpful; we have starred certain parts of the book and placed a wavy rule in the margin to indicate that they may be omitted on first reading. These starred passages are undoubtedly more difficult; this may be (as in the case of some of the material of Section 7 of Chapter 2 dealing with odds) because we are applying the mathematics to a rather sophisticated experimental situation, or it may be (as in the case of the whole of Chapter 6) because the mathematical ideas are themselves more advanced. Naturally, we hope each one of our readers will find some attractive items among the starred parts of our book. But the readers who find these parts difficult should not judge themselves harshly!

0.7 THE EFFECT OF CALCULATORS

Hand calculators are a great boon and enable us to avoid much of the tedium associated with elementary arithmetic. Regrettably, our children, in their mathematical education, are not always able to take full advantage of this marvelous new tool, because their elders and (so-called) betters decree that it is somehow good for them to acquire proficiency in the old algorithms, even if the cost is excruciating boredom and a life-long aversion to what they erroneously suppose to be mathematics. If proficiency in the algorithms is going to continue to be the basis of success in the tests, then the algorithms will continue to figure prominently in the curriculum.

Fortunately, you and we are under no such brutal compulsion. The calculator makes its appearance in many parts of this book, wherever arithmetical calculations are involved. But we do more—we use the calculator not only to avoid boring calculations, but also to reinforce the teaching of mathematics. Many possibilities exist to do this—let us illustrate our point with just one example. Suppose we start by pressing, in turn, the buttons $\boxed{5}$, $\boxed{+}$, $\boxed{7}$ and then repeatedly press the $\boxed{=}$ button; most calculators will generate the sequence

$$12, \quad 19, \quad 26, \quad 33, \quad 40, \quad \cdots.$$

Suppose then that we put to the student the question, 'If you continue in this way, will you hit the number 92?' It is a matter of seconds to discover you do not; you hit 89 and then 96. The student may then, rapidly and painlessly, try other examples. Now comes the question 'How could you have decided without actually pushing buttons on the calculator?' This discussion should lead naturally into an understanding of the idea of remainders and, from there, of modular arithmetic—usually regarded as a pretty advanced topic. Thus the calculator appears as a teaching aid, and the *actual* technique employed to solve a problem first attacked with the help of a calculator may well not involve the calculator at all! Thus *intelligent* use of the calculator should not lead to your using it every time you're faced with a mathematical problem—it may not be the easiest way.

There are, however, three warnings to be given about using the calculator. First, there is a danger of treating the calculator as a sort of god—or, at least, an oracle—and believing that something must be so because the calculator says so. Take, as an example, the product of two negative numbers. Almost all calculators can do calculations with negative numbers, so we can use the calculator to 'discover' that $(-7) \times (-8) = 56$. We have observed students believing that this is a proof that the product of two negative numbers is positive! Of course it is not, and accepting the calculator's answer, without understanding why it has been decided to *define* the product of two negative numbers to be positive, is to accept that same authoritarianism against which we have been arguing so strongly. In this text we go to great trouble and length to make out a convincing case for *human beings* defining the product of two negative numbers in the way in which *human beings* have built it into the calculator circuitry.

Second, calculator arithmetic is not the same as human arithmetic. One important respect in which this is so is that calculators (or, at any rate, most calculators) can't handle fractions and some problems are best tackled with the aid of the arithmetic of fractions. Another important respect in which the two differ is that the calculator is subject to 'round-off error'. Thus, called upon to divide 1 by 3, the calculator shows 0.3333333. If we take this number and multiply it by 3, the result is 0.9999999. Thus multiplying by 3 has not quite succeeded in 'undoing' the result of dividing by 3 on the calculator, as, of course, it would do in purely human arithmetic. We have to be careful because round-off errors can accumulate and become increasingly serious—if we raise 0.9999999 to the 10th power on the calculator the result is 0.9999990.

Third, calculations on the hand calculator should be checked. This is because we may punch the wrong keys, or the calculator may perhaps be defective (many malfunction when the batteries are low). Thus the use of the hand calculator increases the importance of a good understanding of estimation and approximation, the topic of Chapter 4 of this volume. Notice that we check an accurate calculation by a quick, inaccurate one, to be sure that the answer the calculator gives us is sensible.† We do not check the calculator by doing a long, tedious accurate calculation without the use of the calculator (although some elementary texts seem to suggest there is merit in so silly a procedure!).

Is there then no place for the traditional algorithms in a well-planned course of mathematical instruction? We believe they do have a place, but one very different from that traditionally accorded to them. It is important to understand what an algorithm is and to know that there are algorithmic processes. Indeed, if one attempts a definition of *algorithm,* one is led to the notion of a process which can be carried out by a machine, so the very definition itself, together with a humane approach to students, leads to the conclusion that they should not be a regular feature of human activity! So it is important to know what an algorithm is, and to understand why it works. As a practical tool it is best regarded only as an ultimate 'back-up', available if all else fails. And excellent experience in genuine mathematical thinking is to be had in seeing how to *avoid*

---

† Notice, too, that we need to have an absolute mastery of our 'number facts' to make these checks really quick and efficient.

the use of an algorithm in doing a calculation that appears at first sight to require it (in the absence of a calculator). Let us give an example. Suppose we are to multiply 72×68. Now there is an important identity, relating the various arithmetical operations, which asserts that

$$X^2 - Y^2 = (X + Y) \times (X - Y).$$

This may be 'proved by examples', just as we explained in item (d) of Section 6 of this Introduction. Suppose you grant us the truth of this formula, called the formula for *the difference of two squares*. Then, in particular,

$$72 \times 68 = (70 + 2) \times (70 - 2) = 70^2 - 2^2.$$

Now, $7^2 = 49$, so $70^2 = 4900$ and $2^2 = 4$. Thus $72 \times 68 = 4900 - 4 = 4896$. We assure you that, for a person knowing the formula for the difference of two squares, it is a matter of a few seconds to calculate 72×68 in one's head.

But let no one come up with the hoary old argument that you must be proficient in the algorithms in case your hand calculator isn't available.† This is hypocritical rubbish! If you don't have a hand calculator with you, borrow one. If the battery is dead, buy a new one. If the calculator is irreparably damaged, buy a new one. And if you can't apply any of these remedies immediately, wait until you can! Until you find yourself in the situation of unexpectedly having to do an urgent and complicated calculation—something that's never happened to either of us!—don't fall back on a dreary algorithm!

A final word must be said. You're probably having to learn more mathematics in response to the challenge of an advancing technology; and one of the pieces of evidence of that advancing technology is the availability of reasonably priced hand calculators. But as technology advances, so do calculators improve, and their construction and utilization change. Thus you should not get into the habit of depending on one particular model; you should learn flexibility in being able to use several. Fortunately, the key to being able to adapt to changing technology is provided by the very subject you are learning—mathematics!

0.8 SUBSEQUENT VOLUMES

It is our intention to guide the adult through the mathematics he or she is likely to need to get a job, hold a job, to advance a career, and to function effectively in the modern world. Our starting point, in

---

†A car with a dead battery is a far greater inconvenience than a hand-calculator with a dead battery. Yet it is rare for those who advocate drill in dreary algorithms to keep a horse 'just in case'.

this volume, has been determined by our perception, and the judgments of our friends and colleagues, of the level of mathematical mastery of a large number of today's intelligent adults; our terminal point, at the end of Volume 3, will be determined by our intentions for the whole enterprise, as expressed in the opening sentence of this section. Thus we do *not* plan to present all the mathematics needed by the engineer or the scientist (here we refer to the 'hard' sciences—physics, chemistry, biology—rather than the social sciences); but we do believe that such a person will have been prepared, by a study of the content of these volumes, to embark on the more specialized courses appropriate to those disciplines.

There is an English phrase, 'Never buy a pig in a poke'. A poke is a bag or pouch, and the advice is not to launch yourself on a project without knowing what you're taking on! Thus we feel we have a duty to our readers to tell them what we intend to cover in the two subsequent volumes, to achieve our stated objective. In Volume 2 we will launch ourselves more systematically on a study of algebra and geometry. But these subjects will not be taught as separate, unrelated disciplines. In both, we will be stressing the importance of *pictorial representation* (through graphs, figures, etc.), and the study of several topics—for example, the solution of equations—will have explicit algebraic *and* geometric aspects. The notion of a *function* will be strongly featured as one of the key notions of mathematics, embodying the familiar and important idea of one thing changing in response to changes in something else (as the heights of children change as they grow older). With this emphasis, we hope to make much of pre-college algebra far more dynamic and relevant, and less static—and stagnant! We will develop geometry in three dimensions as well as two, and we will pay special attention to the fascinating patterns which can be constructed in three dimensions.

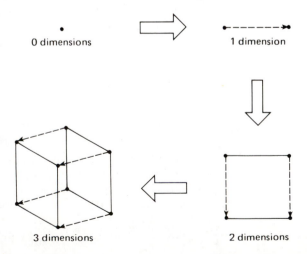

Many of our examples and applications of algebraic technique and geometrical insight (in the broad sense) will be drawn from probability and statistics, so that these topics will themselves receive considerable attention. We regard it as very important to do this, since people are so heavily bombarded, in their daily lives, with statistics intended to convince them of the good sense of some course of action, all the way from changing one's deodorant to installing thermopane glass in one's house; and few of us have the ability to judge the validity of the arguments used.

Volume 2 will contain some trigonometry as a natural outgrowth of algebra and geometry. Our treatment will be based–as any sound treatment must be–on the theory of similar triangles and the famous theorem of Pythagoras, saying that, in a right-angled triangle *ABC,*

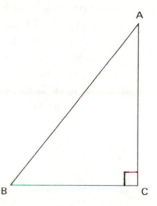

the square of the length of *AB* is the sum of the square of the length of *AC* and the square of the length of *BC.* However, our emphasis in the development we will give will be oriented toward the practical applications of trigonometry, for example, to indirect measurement.

We will not attempt to be comprehensive in Volume 2, any more than we have been in Volume 1. It is not part of our plan to set down everything which is known about the topics we treat. We will be content if we make our readers comfortable with those topics–if they feel the need (or desire) for further reading they will be able to consult the bibliography we will provide.

It would not be sensible to set down in any detail the intended content of Volume 3, since the reader needing Volume 1 would not be expected to be in a position to appreciate such a description. But the key word for Volume 3 is–*the calculus.* It will be our objective to describe what the calculus is about–the type of problems it is designed to solve and the way it solves them. We will deal with the two complementary aspects of the calculus–first, the way it can be used to study rates of change (the *differential* calculus), and, second, the way it can be used to study global effects of change (the *integral* calculus). Remember that readers of Volume 1 will already be in possession of good notation and terminology; and readers of Volume 2 will already have become thoroughly familiar with the function concept and pictorial representation of functional relations through graphing. Thus we have solid grounds for supposing that the introduction of the fundamental notions of the calculus will not present our readers with the problems that so many have found at this stage of their mathematical education.

In the subsequent volumes we will continue to adopt the relaxed, informal style of the present volume. To be sure, we will present genuine mathematical reasoning, and this reasoning will, in a sense, be subtler than that encountered in this volume. But it should not strike the reader as so; for what we notice is change, and the level of mathematical reasoning (which we took to be zero, for convenience, at the outset of Volume 1) should have been steadily rising throughout the study of the three volumes. Our objective will have been achieved if the reader finds himself or herself having a smooth, unbumpy ride through our text and asks at the end 'Is that all?' Then we will feel that we have achieved our intention—indeed, our sincerest ambition—to provide a guide through mathematics for you, the intelligent adult.

1

Decimals

Those not entirely comfortable with whole-number algorithms may wish to read the appendix before starting this chapter.

1.1 DOLLARS AND CENTS

We know that 382 cents has the same value as 3 dollars and 82 cents. We write the former as 382¢ and the latter as $3.82, so we may also write

$$382¢ = \$3.82$$

Obviously, when doing arithmetic with money it makes no difference whether we work just with cents or with dollars and cents. If we have to add 382¢ and 1247¢ we do the addition

```
   382
+ 1247
  1629
```

and deduce that the answer is 1629¢. We can use the *same* addition to infer that if we add $3.82 and $12.47 we get $16.29. If we wanted to retain, in our addition, the fact that we are working in dollars and cents, instead of just in cents, we could 'put in the dots' and write

```
   3.82
+ 12.47
  16.29
```

and so deduce that the answer is $16.29.

Exactly the same rule applies to subtraction. If we want to subtract 417¢ from 978¢ we may do the subtraction

$$
\begin{array}{r}
978 \\
-417 \\
\hline
561
\end{array}
$$

inferring the answer 561¢; or we may express the quantities in dollars and cents, put in the dots, *do the same subtraction,* now writing it

$$
\begin{array}{r}
9.78 \\
-4.17 \\
\hline
5.61
\end{array}
$$

and infer the answer $5.61.

Finally, the same principle applies to multiplication by a whole number. Suppose your weekly take-home pay is $321.75; how much do you make in 4 weeks? You can work in cents so that you have to carry out the following multiplication.

$$
\begin{array}{r}
32175 \\
\times \quad\quad 4 \\
\hline
128700
\end{array}
$$

Alternatively you can work in dollars and cents, put in the dots, *do the same multiplication*

$$
\begin{array}{r}
321.75 \\
\times \quad\quad 4 \\
\hline
1287.00
\end{array}
$$

and infer that in 4 weeks you earn $1287.00.

The principle then is this: when adding amounts of money, subtracting amounts of money, or multiplying amounts of money by whole numbers, we may either work just in cents or in dollars and cents. The arithmetic, in the two cases, is exactly the same; but, if we work in dollars and cents, we put in a dot to mark where the dollars 'end' and the cents 'begin'.

Remarks

(i) We could write in dollars and dimes, but this is not customary. If we did, then it would be reasonable to represent, say, 17 dollars and 4 dimes as $17.4. The arithmetic of dollars and dimes would then be very much like the arithmetic of cents or of dollars and cents. That is, $17.4 means exactly the same as 174 dimes. If we invent the notation ₡ for dimes, then we may write

$17.4 = 174₡.

We can go further; for 174₡ is the same as 1740¢ so that

$17.4 = 174₡ = 1740¢ = $17.40.

	1940	1980
	12¢	11d
	16¢	15d
	5¢	5d
	400¢	400d

Is our dime now equivalent to the old cent?

Thus, looking at the beginning and the end of the string of equations above, we have

$17.4 = $17.40.

In other words, we may put a zero at the extreme right of the symbol $17.4 without changing its real meaning. The only difference is that between talking of '4 dimes' and talking of '40 cents'. Then, as we have said, we do ordinary arithmetic (addition, subtraction, multiplication by whole numbers) with dimes and 'put in the dot', if we wish, to represent the sums of money involved in dollars and dimes. The point goes in, in this case, one place from the right-hand end. In dealing with dollars and cents, however, the point goes in *two* places from the right-hand end.

(ii) In one of the problems discussed above, we came up with the answer $1287.00. It would, of course, be perfectly legitimate then to suppress the '.00' in this symbol and simply announce the answer $1287, since

$1287.00 = $1287

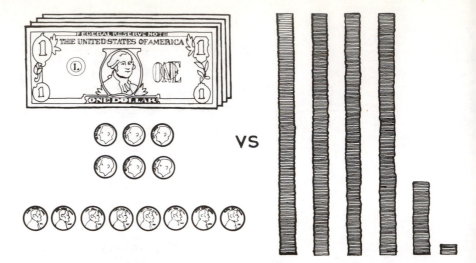

However, there *could* be circumstances in which the answer '$1287.00' to a question gave more precise information than the answer '$1287'. If we ask for a person's monthly salary and are given the answer '$1287.00', we would know that the figure is given *precisely*; but if we are given the answer '$1287', we might well think that the figure is being given to the nearest dollar, or with the number of cents being ignored. Thus we may include the number of cents, *even if there are no cents,* to indicate the accuracy of the figure being quoted.

(iii) The word 'deci' comes from the Latin word *decimus,* meaning 'tenth'. One cent is a tenth of a dime, one dime is a tenth of a dollar, etc.; so when we put in the dot to mark the separation between either dollars and dimes or dollars and cents, the calculations may then be conveniently referred to as *decimal* arithmetic.

(iv) We can correctly write

$4.68 = $4 and 6¢ and 8¢

and

$4.68 = 400¢ and 60¢ and 8¢.

The expressions on the right in the two cases certainly represent the same amount of money but there is a subtle difference in the way we think of them. It is natural to think of the first as modeling a real-life situation involving four dollar bills, six dimes and eight pennies, while the second would then model a real-life situation involving 468 pennies. However, the symbols we write down to express our arithmetic computations are not always in one-to-one correspondence with the actual objects they represent. As a practical matter this poses no real problem since, for example, we are quite accustomed to accepting two nickels as equivalent in value to a dime, if a cashier has no dimes. (We might protest for other reasons, however, if offered two thousand pennies instead of a twenty dollar bill.)

Exercise 1.1

1. In the symbol '$28.31' the '1' represents '1 cent' and the '3' represents '3 dimes' (or '30 cents').

   (a) What does the '8' represent?

   (b) What does the '2' represent?

*2. (a) Explain why it would not be convenient to make a model of decimal arithmetic using nickels.

   (b) What other denominations of coins are unsuitable for making a model of decimal arithmetic?

3. What is the smallest number of bills and coins you could have in your pocket that would have a value of four dollars and sixty-eight cents?

4. **A Puzzle Question** (pure fun)

   We have observed that

An amount of money equivalent to the following number of cents	can be obtained with the following number of coins (or bills)	and an example of a specific collection of coins that will work is the following.
$1 \times 1 \times 1 = 1$	$1 \times 1 = 1$	1 penny
$2 \times 2 \times 2 = 8$	$2 \times 2 = 4$	1 nickel 3 pennies
$3 \times 3 \times 3 = 27$	$3 \times 3 = 9$	2 dimes 7 pennies
$4 \times 4 \times 4 = 64$	$4 \times 4 = 16$	12 nickels 4 pennies
$5 \times 5 \times 5 = 125$	$5 \times 5 = 25$	?

How far can you extend this list?

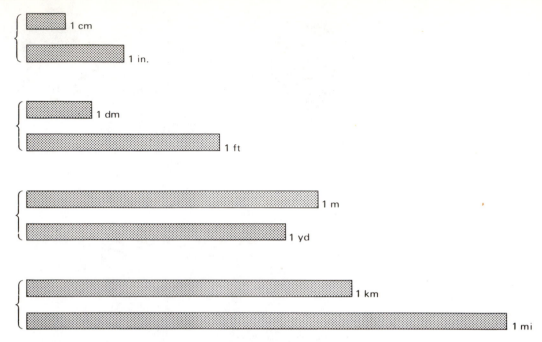

Relative lengths of metric and American units

1.2 METRIC MEASURES OF DISTANCE AND WEIGHT

In our American system of measuring distances, we measure relatively long distances in miles (for example, distances between cities), shorter distances in yards and feet, and even shorter distances in feet and inches. Elsewhere in the world kilometers are used for measuring distances between cities, and shorter distances are measured in meters or centimeters; sometimes a measure between the meter and the centimeter, called the 'decimeter', is also used. This system is called the *metric* system. You can get an idea of metric lengths from the following approximate conversion table between the two systems:

8 kilometers   is almost the same as   5 miles.
10 meters   is almost the same as   11 yards.

However, the great advantage of the metric system lies in the very simple way we convert measures within the metric system. Compare these two tables:

American measures of distance	*Metric measures of distance*
1 mile = 1760 yards	1 kilometer = 1000 meters
1 yard = 3 feet	1 meter = 10 decimeters
1 foot = 12 inches	1 decimeter = 10 centimeters

Which is easier!? Surely there can be no doubt that the metric system is far easier. It is not only easier to remember the conversion; it is also far easier to do arithmetic with metric measures than with the American measures. This is because, as you can see from the two tables above, metric measures use the number base 10, our standard base for writing numbers, while the American system uses no number base at all!

Let us suppose we are dealing with distances measured in meters and centimeters. We know that there are 100 centimeters in a meter, so that a distance of 3 meters and 82 centimeters is the same as a distance of 382 centimeters. Guided by our notation for dollars and cents, we might write this

$$382 \text{ cm} = \text{m } 3.82$$

(Actually it is customary to write the 'm', for 'meter', like the 'cm', for 'centimeter', on the right, but let us adopt the convention above temporarily). You will notice that this formula looks remarkably like the first formula of Section 1.1. Of course it does, because the relation of centimeters to meters (1 meter = 100 centimeters) is *exactly* the same as the relation of cents to dollars (1 dollar = 100 cents). Thus the addition sum

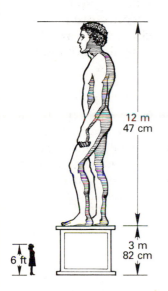

12 m
47 cm

3 m
82 cm

6 ft

$$
\begin{array}{r}
382 \\
+ 1247 \\
\hline
1629
\end{array}
$$

not only tells us that if we add together 382¢ and 1247¢ we get 1629¢, but also that if we add together 382 cm and 1247 cm we get 1629 cm. Even more important, the *same* addition sum, in the version

$$
\begin{array}{r}
3.82 \\
+ 12.47 \\
\hline
16.29
\end{array}
$$

not only tells us that if we add together \$3.82 and \$12.47 we get \$16.29, but also that if we add together m 3.82 and m 12.47 we get m 16.29. That is, if we put a piece of sculpture 12 meters, 47 centimeters high on a platform 3 meters, 82 centimeters high, the sculpture will rise 16 meters, 29 centimeters above the ground. The *same* calculation serves for problems involving money and problems involving distance (or length, height, etc.). Notice that this wouldn't be true at all with the American system. If one fence is 12 yards and 21 inches long and another is 7 yards and 15 inches long, together they are 20 yards long. But if one person has 12 dollars and 21 cents and another has 7 dollars and 15 cents, it is *not* true that together they have 20 dollars—they only have 19 dollars and 36 cents.

We can extend our remarks on the connection between the arithmetic of money and the arithmetic of metric measures of distance (using meters and centimeters) to include addition, subtraction and multiplication by whole numbers. The arithmetic is, in both cases, exactly the same—and it is the ordinary arithmetic of the whole numbers, except for 'putting in the dot' to mark the separation between dollars and cents, or between meters and centimeters.

If you go back to remark (i) at the end of Section 1.1, you should be able to see the connection between dealing with money in terms of dollars and *dimes*, and dealing with distances in terms of meters and *decimeters*. Again, the arithmetic is exactly the same in the two cases.

Now suppose we are considering greater distances, so that it is natural to use the kilometer measure. We may record a distance as 7 kilometers, 583 meters. This is the same as 7583 meters. What could be more natural than that we should use the 'dot' notation again and write

7583 m = 7.583 km.

(Here we follow the standard practice of writing 'm' and 'km' to the *right* of the number; you may write them on the left if you feel more comfortable doing so.) Again we can carry out calculations either in meters or in kilometers and meters, and the arithmetic is the same in the two cases. Thus if you jog 4,129 meters,† take a short rest, and then jog 2,661 meters, you can work out how far you have jogged by doing this addition.

$$\begin{array}{r} 4,129 \\ +\ 2,661 \\ \hline 6,790 \end{array}$$

But if we want to record this total distance in kilometers and meters, we would write

$$\begin{array}{r} 4.129 \\ +\ 2.661 \\ \hline 6.790 \end{array}$$

and we would know that the total distance jogged was 6 kilometers, 790 meters. Once again, you should understand that you can also subtract distances, measured in kilometers and meters, and multiply such distances by whole numbers, by using ordinary arithmetic and putting in the dots where they should go.

We come now to a second, tremendous advantage of the metric system over the American system; all that we have said about distance

4,129+2,661=?

†We often separate the digits of a large whole number into groups of three by means of commas, starting from the right, to make the number easier to read. Thus we write 4,129 instead of 4129, and 1,000,000,000 instead of 1000000000.

applies equally well to weight (and, indeed, to measures of capacity—based on the liter—too). Just as distance in the metric system is based on the meter, weight is based on the gram and we have *exactly the same* conversion table within the metric system of weights as we had for distances.

Metric Measures of Weight

1 kilogram	=	1000 grams
1 gram	=	10 decigrams
1 decigram	=	10 centigrams
1 centigram	=	10 milligrams

You will notice that we have an extra item in our table here, compared with the table for distances, namely, the last item giving the conversion between centigrams and milligrams. But, as you might guess, we *could* have put millimeters into our table of distances and it *is* true that

1 centimeter = 10 millimeters.

(You may be more familiar with the fact that 1 meter = 1000 millimeters, just as 1 gram = 1000 milligrams.) We only postponed mentioning this so that we should have the same number of items on both sides when comparing American and metric measures of distance.

The American system of weights is based on the pound and the approximate conversion rule between the system is

10 kilograms is almost the same as 22 pounds.

But you will now see that the American system of weights and measures isn't a 'system' at all in the sense that the metric system is a real 'system'! When we say that the meter is the base for the metric measure of distance, we mean that the other units are the *kilo*meter ('kilo' is Greek for 1000), the *deci*meter ('deci' is Latin for 10), the *centi*meter ('centi' is Latin for 100), and the *milli*meter ('milli' is Latin for 1000). It is in exactly the same sense that the *gram* is the base for the metric measure of weight, and that the *liter* is the base for the metric measure of volume. In what sense can we say that the foot, the pound, the gallon are the bases for the American measures of distance, weight, volume?

You will now see that all that we have said about the arithmetic of kilometers and meters applies without change to the arithmetic of kilograms and grams. We hope that you also see it applies to the arithmetic of meters and millimeters and to the arithmetic of grams and milligrams. The important aspect is simply that the larger unit (kilometer, kilogram, meter, gram) should be 1000 times the smaller unit (meter, gram, millimeter, milligram). The subtraction

$$
\begin{array}{r}
12.108 \\
-\ \ 7.453 \\
\hline
4.655
\end{array}
$$

tells us that

(a)   12 kilometers, 108 meters less 7 kilometers, 453 meters is 4 kilometers, 655 meters,

(b)   12 kilograms, 108 grams less 7 kilograms, 453 grams is 4 kilograms, 655 grams,

(c)   12 meters, 108 millimeters less 7 meters, 453 millimeters is 4 meters, 655 millimeters,

(d)   12 grams, 108 milligrams less 7 grams, 453 milligrams is 4 grams, 655 milligrams,

—and a host of other facts, too!

Remarks

(i) Our second remark relating to money (is $1287 the same as $1287.00?) holds even more strongly in the case of reported measurements. If we ask you the length of a room and you report 3.680 meters, we will infer that you have measured the length of the room to the nearest millimeter. But if you report 3.68 meters, we will infer that you have only measured the length of the room to the nearest centimeter. Thus the number of digits on the right of the dot indicates the accuracy claimed for the measurement. Of course, it is strictly true that 680 millimeters = 68 centimeters; but, in practice, we would only refer to millimeters if our measuring instruments, and the nature of the problem, justified us in referring to lengths to the nearest millimeter.

(ii) *Meter vs metre.*   It is reported that the United States agreed to adopt the spelling 'metre' provided that Great Britain adopted the spelling 'gram' rather than 'gramme'. Certainly the spelling 'gram' is used in Great Britain, but the United States has stuck to the spelling 'meter', rather than 'metre'. This has the effect that the United States is out of line with the rest of the English-speaking world. However, this kind of uniqueness is not new—the spellings 'theater' and 'specter' exhibit the same rugged individualism. (So, too, does the spelling 'liter'.)

But it can be argued that this special spelling of 'meter' is harder to justify, since the great merit of the metric system is that it is genuinely international. It is also arguable that we use the term 'meter' in various compound words to refer to measuring instruments ('speedometer', 'Odometer', 'light meter'), so that it would have been less confusing to use 'metre' for a very specific length.

We have felt that we have no choice but to adopt the United States spelling!

Exercise 1.2 (For practice and appreciation)

In Tables I and II shown below the numbers along horizontal rows represent measurements of distances. In each case we can read these measurements in various ways and we can also infer, from the data, the accuracy of the recorded measurement.

Table I

	Kilometers	Hectometers	Dekameters	Meters	Decimeters	Centimeters	Millimeters
(a)		8	6	2	1	0	
(b)		7	6	4	1	2	0
(c)	1	7					
(d)	1	0	2	6	7	8	
(e)	4	0	7	2	3		
(f)		9	1	7	3	4	

10 millimeters = 1 centimeter
10 centimeters = 1 decimeter
10 decimeters   = 1 meter
10 meters   = 1 dekameter
10 dekameters   = 1 hectometer
10 hectometers = 1 kilometer

Table II

	Miles	Yards	Feet	Inches
(a)		8	2	11
(b)		14	2	31
(c)	1	0	9	18
(d)		6	2	
(e)		1759	2	11
(f)		1758	3	36

12 inches   = 1 foot
3 feet   = 1 yard
1760 yards = 1 mile

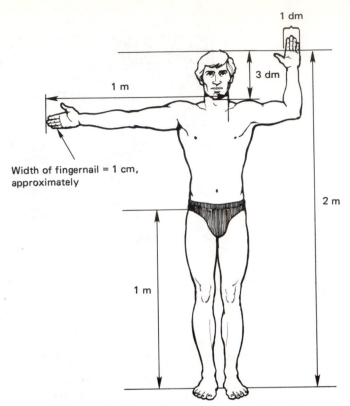

Approximate metric body dimensions

For example, from the data in row (a) of Table I we can assume the measurement to be accurate to the nearest centimeter and the measurement may be expressed in any of the following ways:

86210. centimeters (or 86210 centimeters)
8621.0 decimeters
862.10 meters
86.210 dekameters
8.6210 hectometers
.86210 kilometers

Notice how the dot's position relates to the name of the unit.

We could also write 862100 millimeters, but this would imply more accuracy than is justified by the data; so we would not ordinarily choose this form. Also, in practice the dekameter and hectometer forms are seldom used. However, they are not wrong and you should use the units that are most convenient and natural for your purpose.

The data in Table II can be interpreted in a similar way. Thus, in row (a) of that table we can infer that the measurement is accurate to the nearest inch and the measurement may be expressed in any of the following ways:

323 inches
26 feet and 11 inches (or about 26.92 feet)
8 yards and 2 feet and 11 inches (or about 8.97 yards)

1. Choose either Table I or Table II and, for the entries (b) through (f) in the table of your choice (do not do both) state how accurate the measurement is likely to be and express the distance in three different ways.

2. Certain manufacturers are required by law to state the minimum weight of the items they sell. The result is that a loaf of bread, for example, has been seen with the label 'NET WT. 16 oz (= 453.592 grams)'. Is this a 'sensible' way for the manufacturer to label the weight of an item? (Consider the degree of accuracy claimed for '16 oz' and for '453.592 grams'.) Next time you are in a store look for items that are labeled 'unreasonably'.

3. Notice that we have not discussed *converting* measurements in one system to equivalent measurements in other systems. This is because in real-life situations one would seldom be called on to do this. However, it is useful (psychologically) to have a concrete feel for the units of measurement you use. Find a ruler (note we did not commit the common error of calling this a yard stick) calibrated in centimeters and measure several familiar objects, so that you will get an intuitive feel for various lengths in the metric system.

4. Find a gradeschool textbook in mathematics and see how many absurd problems you can find regarding measurement (for example in some old textbooks you may see: 'Change 2 feet, 3 inches to meters.'). Perhaps this will help you to see one reason why so many people feel unnecessarily anxious about mathematics.

5. **A Puzzle Question (for fun)**

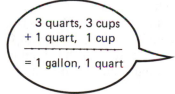

3 quarts, 3 cups
+ 1 quart, 1 cup
———————
= 1 gallon, 1 quart

59 minutes, 59 seconds
+   ? minutes,   ? seconds
———————
= 1 hour, 1 minute

gallons	quarts	cups
	3	3
+	1	1
1	1	0

Case I

hours	minutes	seconds
	59	59
+	?	?
1	1	0

Case II

quarters	nickels	pennies
	4	4
+	?	?
?	?	?

Case III

dekaliters	liters	deciliters	centiliters
	9	9	9
+	?	?	?
?	?	?	?

Case IV

(a) Fill in the appropriate values for '?' in Case II.

(b) Complete Cases III and IV so that they are 'like' Cases I and II.

(c)   Find units that would make this computation valid.

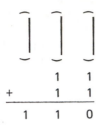

$$
\begin{array}{ccc}
  & 1 & 1 \\
+ & 1 & 1 \\
\hline
1 & 1 & 0
\end{array}
$$

1.3 DECIMALS

As you will surely by now be convinced, calculations of the kind

$$
\begin{array}{ccc}
\begin{array}{r} 3.82 \\ +\ 12.47 \\ \hline 16.29 \end{array}
&
\begin{array}{r} 12.108 \\ -\ 7.453 \\ \hline 4.655 \end{array}
&
\begin{array}{r} 321.75 \\ \times\quad\ \ 4 \\ \hline 1287.00 \end{array}
\end{array}
$$

which are really just ordinary arithmetic 'with the dot put in', have a place in a wide variety of applications of mathematics. It would therefore be natural to extend our idea of number so that, for example, 3.82, 12.47, and 16.29 become numbers. Likewise the definitions of addition, subtraction and multiplication by a whole number can be extended so that it becomes true that, for example, 3.82 + 12.47 = 16.29 *as numbers*. After all, this is exactly what we do at the very beginning of arithmetic, when we *abstract* from such experiences as that of finding that if 3 apples are put into a bag containing 5 apples there are then 8 apples in the bag, to obtain the purely numerical statement 5 + 3 = 8. Thus, just as in the elementary case, we certainly want our idea of number and our definitions to be such that the statement

$$3.82 + 12.47 \ = \ 16.29$$

should generalize the facts that $3.82 + $12.47 = $16.29 and 3.82 m + 12.47 m = 16.29 m.

The clue as to how to do this is contained in our statement on page 000 that the same methods work for meters-centimeters as for dollars-cents "because the relation of centimeters to meters (1 meter = 100 centimeters) is *exactly* the same as the relation of cents to dollars (1 dollar = 100 cents)". Now since 1 dollar = 100 cents, it follows that 1 cent is $\frac{1}{100}$ of a dollar, so that 82 cents is $\frac{82}{100}$ of a dollar. Thus $3.82 is $3\frac{82}{100}$ dollars. Similarly, 3.82 m is $3\frac{82}{100}$ meters, and 3.82 grams is $3\frac{82}{100}$ grams.

Again, reverting to Section 1.1 (Remark (i)), we know that 1 dime is $\frac{1}{10}$ of a dollar, so $17.4, which means 17 dollars and 4 dimes, is $17\frac{4}{10}$ dollars. And—one more example—1 meter is $\frac{1}{1000}$ of a kilometer, so 7.583 km = $7\frac{583}{1000}$ km.

These examples lead us to make the following definition.

Definition *Decimal numbers* (or *decimals*) are numbers of the form 17.4, 3.82, 7.583, 12.4267, where a dot is introduced between a pair of adjacent digits of an ordinary whole number. The *meaning* of such a decimal can be explained as follows in terms of fractions (or mixed numbers).† The part to the left of the dot is the whole number part of the mixed number. The part to the right of the dot is the numerator, or top, of the fractional part of the mixed number. The denominator, or bottom, of the fractional part is a power of 10— the first power (10) if there is only one digit to the right of the dot, the second power (100) if there are two digits to the right, the third power (1000) if there are three digits to the right, etc. Thus

$$17.4 = 17\frac{4}{10}, \quad 3.82 = 3\frac{82}{100}, \quad 7.583 = 7\frac{583}{1000},$$

$$12.4267 = 12\frac{4267}{10000}.$$

The dot is usually referred to as the *decimal point*.

Note now that *we have no need to invoke the addition and subtraction of fractions to know how to add and subtract decimals.* We've done all this in Sections 1.1 and 1.2 and know that it's just a matter of ordinary familiar addition and subtraction of whole numbers 'with the dot put in', *provided that both decimals* have the same number of digits to the right of the dot. Suppose they don't; suppose you're asked to work out 12.6 + 3.57. What do you do then? Here the clue is provided by our earlier remark that $17.4 is the same as $17.40, so that we would expect that 17.4 is the same decimal number as 17.40. The general rule is this:

We may put as many zeros as we like to the right of the digits of a decimal number, without changing its numerical value.

Thus, as decimal numbers, 12.28 = 12.2800, 14.1 = 14.1000, and, turning back to our example, 12.6 = 12.60. We now see that we may add 12.6 + 3.57 by rewriting the problem as 12.60 + 3.57. It is displayed as follows

$$\begin{array}{r} 12.60 \\ +\quad 3.57 \\ \hline 16.17 \end{array}$$

---

†We do not require here anything about the arithmetic of fractions. We only suppose that the *idea* of a fraction is familiar to the reader.

and we obtain the answer 16.17. (When you get expert at this, you may leave out the zero, remembering of course to line up the dots in a vertical line.) If you want to see that 12.6 really is the same as 12.60, *as a number,* observe that $\frac{6}{10} = \frac{60}{100}$, that is those are equivalent fractions, so that $12\frac{6}{10} = 12\frac{60}{100}$.

But actually you *shouldn't* be asked to add 12.6 + 3.57! Such problems are usually set as traps for the unwary and do not arise in real-life situations! For, in real-life situations, measurements (or amounts of money) that need to be added or subtracted are always quoted in the same units, so that, in our mathematical model, the decimals that represent these measurements will have the same number of digits after the decimal point. It is unnatural to pose the question 'If the distance from *A* to *B* is 12 meters, 6 decimeters and the distance from *B* to *C* is 3 meters, 57 centimeters, what is the distance from *A* to *C* via *B*?' The *natural* question gives the distance from *A* to *B* as 12 meters, 60 centimeters.

Obviously decimals can be multiplied by whole numbers just as we did in the special case of dollars and cents in Section 1.1. However, a particularly nice situation arises when we multiply by powers of 10. Suppose for example, we have to multiply 173.46 by 10. The standard calculation gives this result.

$$
\begin{array}{r}
173.46 \\
\times \quad 10 \\
\hline
1734.60
\end{array}
$$

Compare the number of places the dot gets shifted with the number of zeros following the '1' in the multiplier.

What has happened? The answer, 1734.60, contains the same sequence of digits as the original number to be multiplied, 173.46, except that (i) a zero has appeared on the extreme right and (ii) the dot now appears between the 4 and the 6, instead of between the 3 and the 4, that is, *the dot has shifted one place to the right.* Now, as we have mentioned above, it is largely a matter of taste whether you leave that final zero in the decimal expression or not. The numerical value of the decimal is not affected. So the essential fact is the shifting of the dot, and it is easy to see that this always happens when you multiply by 10. In just the same way, you can see that multiplying by 1000 shifts the dot three places to the right, etc. We therefore have a very easy rule for multiplying decimals by powers of 10. For example,

$$173.46 \times 10 \quad = 1734.60 = 1734.6,$$
$$173.46 \times 100 \quad = 17346.00 = 17346,$$
$$173.46 \times 1000 = 173460.00 = 173460.$$

Notice that in the second formula above we have no dot at all appearing in 17346, and in the third formula, it is *essential* to keep the zero—we are not here dealing with a decimal but a whole number.

Just as with dollars and cents, we may regard a whole number as a decimal if we like; $1287 = $1287.00, so 1287 = 1287.00. Indeed

$$1287 = 1287.0 = 1287.00 = 1287.000 = \cdots ;$$

as we have said, we may put in as many zeros *after* the decimal point as we wish, without affecting the numerical value. Thus we may regard an ordinary whole number as a decimal if we wish, and the rule for multiplying decimals by powers of 10 then also applies to whole numbers.

We have not discussed division in this chapter at all, since we wish to devote a separate chapter to this topic. However, the special case of division by powers of 10 can easily be treated at this stage of the development of our understanding of decimals. For multiplication by 10 (say) and division by 10 are inverse operations; each *undoes* the effect of the other. Similarly multiplication by 100 (1000, 10000, etc.) and division by 100 (1000, 10000, etc.) are inverse operations. Thus, since multiplication by 10 is effected by shifting the dot one place to the *right,* division by 10 is effected by shifting the dot one place to the *left.* Similarly, division by 100 is effected by shifting the dot two places to the left, etc. For example

$$173.46 \div 10 \quad = 17.346,$$
$$173.46 \div 100 \quad = \quad 1.7346,$$
$$173.46 \div 1000 = \quad .17346.$$

The third formula above exemplifies another notational convention. We may just as well write 0.17346 (or, indeed, 00.17346, etc.) instead of .17346, we do not thereby change the value of the decimal. In fact, many people make the notational rule that one *must* put one zero to the left of the decimal point rather than leave the left of the decimal point blank. It is silly to make such a rule— we should use whichever notation is most convenient for the problem in hand—but plainly it is useful at least to *think* of zeros being written to the left of the decimal point if we are dividing by 10, or a power of 10 (the situation here is, of course, very analogous to that of imagining zeros to the right when multiplying by 10). Thus, for example, .17346 is the same as 0.17346, so that our rule gives

$$.17346 \div 10 = 0.17346 \div 10 = .017346.$$

*   *Remark*
    The notation 348.156 is not perfect from an arithmetical point of view. A superior notation would be

    348156

    with the dot *above* the digit representing the units. Then we can

Multiplication: Push to the right   Division: Push to the left

imagine the dot being moved freely to the right or left above the displayed number as we multiply or divide by powers of 10, and we can further imagine as many zeros to the left and right of the displayed number as we may care to write. There is the psychological advantage that the dot slides smoothly without having to 'get round' the digits in its path. But there is the further advantage that the notation is then *symmetric,* to the left and right of the dot, in the following sense.

etc.	Two places to the left of the dot marks the hundreds.	One place to the left of the dot marks the tens.	The dot is above the units.	One place to the right of the dot marks the tenths.	Two places to the right of the dot marks the hundredths.	Three places to the right of the dot marks the thousandths.	etc.
	3	4	· 8	1	5	6	

Notice that this is NOT TRUE of the standard notation 348.156. You should write down for yourself the similar statements which *are true* for the standard notation.

Exercise 1.3

1. (Thinking ahead) Sometimes looking over an entire situation before you begin to solve a problem can save you time and effort. Carefully look over the following set of problems, and then compute the answers.

    (a) $3.40 + $7.86 =
    (b)   14.83 m + 9.70 m =
    (c)   57.21 grams + 42.79 grams =
    (d)   572.1 cm + 427.9 cm =
    (e)   3.40 m + 7.86 m =
    (f)   4167 + 3074 =
    (g)   14.83 liters + 9.70 liters =
    (h) $57.21 + $42.79 =
    (i)   1483¢ + 970¢ =
    (j) $41.67 + $30.74 =
    (k)   3.40 grams + 7.86 grams =
    (l)   5721 cm + 4279 cm =
    (m) $14.83 + $9.70 =
    (n)   416.7 decimeters + 307.4 decimeters =

2.   (Some practice and patterns)   Use the rules discussed in this section for multiplying and dividing decimal numbers by powers of 10 and write the answers to the following sets of problems. Where a part of a problem in a set is missing guess what that part should be to fit the pattern and then complete the problem using your guess.

(a) $431 \times 10 =$
$431 \times 100 =$
$431 \times ? =$
$431 \times 10000 =$

(b) $? \times 10 =$
$87 \times 100 =$
$76 \times 1000 =$
$65 \times 10000 =$

(c) $98.76 \times 100 =$
$87.65 \times 1000 =$
$76.54 \times 10000 =$
$? \times 100000 =$

(d) $43 \div 10 =$
$54.3 \div 100 =$
$65.43 \div 1000 =$
$? \div 10000 =$

(e) $.12 \div 10 =$
$2.34 \div 100 =$
$? .56 \div 1000 =$
$456.78 \div 10000 =$

1.4   MULTIPLICATION OF DECIMALS

As we have emphasized so many times, the arithmetic of decimals, if confined to the addition and subtraction of decimals, and the multiplication of decimals by whole numbers, is just the ordinary arithmetic of whole numbers 'with the dot put in'. However, a new element enters when it comes to multiplying one decimal by another. For here we are in the situation where we first have to give a *meaning* to the product of two decimals. If we know what it means to add decimals (as we do!) we already *know* what it means to multiply a decimal by a whole number, since multiplication by a whole number is just repeated addition.

$$4 \times 17.38 = 17.38 + 17.38 + 17.38 + 17.38$$

But this does not tell us what *meaning* to assign to, say, 4.12×17.38, since multiplication by 4.12 cannot be interpreted as repeated addition—it makes no sense to add 17.38 to itself 4.12 times! Thus, before we can say how we should carry out the multiplication of one decimal by another we had better decide what such a product should mean. It could happen (in fact, it doesn't!) that we would find no useful meaning for the product of two decimals, in which case we could save ourselves a lot of useless effort.

Thus we look for situations in which it might be useful to think in terms of multiplying decimals. We should, of course be guided in this search by one very important consideration: if we do come up with a rule for multiplying together two decimals then, if either of them is a whole number, our rule should give us the familiar result of multiplying a decimal by a whole number. Whole numbers, as explained in the previous section, are special decimals, and our rule for multiplying decimals, if we find one, must generalize the known rule for multiplying a decimal by a whole number and not conflict with it.

So let us look at our examples in Sections 1.1 and 1.2. The example of money, expressed in dollars and cents, is not immediately helpful. Can you imagine a situation in which you would wish to multiply together two amounts of money? Surely not! You may, if you're fairly ingenious, think of situations in which you'd like to multiply a given amount of money, say $17.38 by the decimal number 4.12 (we'll have more to say about this when we discuss fractions in a later chapter), but there would in any case be a disadvantage in this procedure, since the two numbers 17.38 and 4.12 are obviously going to play very different roles in this model.

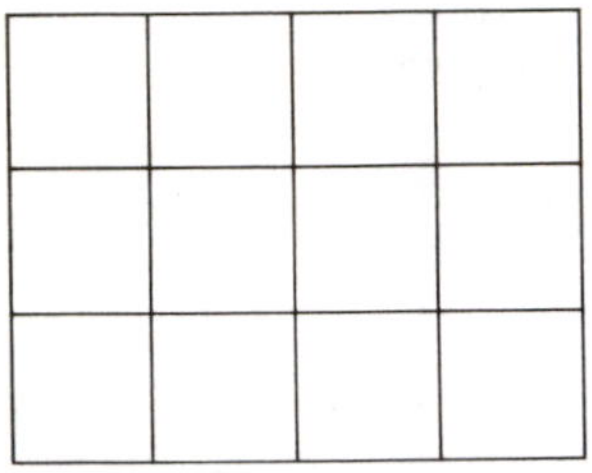

Let us then move on to Section 1.2 and the discussion there of distances measured in the metric system. Do we ever want to multiply together two distances? YES! We multiply distances when we wish to compute the *area* of a rectangle. If the length of a rectangle is 4 cm and the width is 3 cm, its area is 12 square centimeters, denoted 12 sq cm (or 12 cm^2). The geometric interpretation of this fact is seen in Figure 1. Here then is the clue we have been seeking. If the length of a rectangle is 3.12 m and the width is 2.28 m, then the product of 3.12 × 2.28 must be so defined that it gives the area of the rectangle in square meters and square centimeters. What is, in fact the area of our rectangle? Well, its length is 312 cm and its width is 228 cm so that its area is 312 × 228 = 71136, in square centimeters. Now how many square centimeters (cm^2) are there in a square meter (m^2)? We can easily answer this: 1 m = 100 cm, so 1 m^2, which is the area of a square of side 1 m, is also the area of a square of side 100 cm, or (100 × 100)cm^2, which is 10000 cm^2. We write this result as

$$1 \text{ m}^2 = 10000 \text{ cm}^2.$$

Thus, 71136 cm^2 = (70000 cm^2 + 1136 cm^2) = (7 m^2 + 1136 cm^2) = 7.1136 m^2. We have been led to the result

$$3.12 \times 2.28 = 7.1136.$$

This result has some very attractive features. (We hope you agree!) First, it gives a result which looks very reasonable. Since 3 × 2 = 6, 4 × 3 = 12, the product 3.12 × 2.28 should surely be between 6 and 12, and probably nearer to 6. Second, our method of getting it preserves the symmetry of the roles of the two numbers 3.12 and 2.28; we would obviously obtain, by the same argument

$$2.28 \times 3.12 = 7.1136.$$

The third feature is that our rule does conform with what we want (and need!) when one of the numbers in the product is a whole

Figure 1

number. Suppose now that we seek the area of a rectangle 3 m by 2.28 m. Proceeding as above, we convert to centimeters so that the area is (300 × 228) cm², or 68400 cm². But this is just 60000 cm² + 8400 cm², which is 6.8400 m², or 6.84 m². By this method

$$3 \times 2.28 = 6.84,$$

which is exactly what we already knew should be the answer. Thus we plainly have a good and useful rule here, and it remains only to formulate it precisely. The rule says: given any two decimals, each having 2 digits to the right of the dot, we may multiply them by first forgetting the dots and multiplying them as whole numbers, and then putting in a dot so that 4 digits appear to the right of the dot. You may check the examples to see that this is precisely what we do.

It now only remains to see how to generalize this to two arbitrary decimals. Suppose we want 2.141 × 3.622. We may think of a calculation involving meters and millimeters. It should become clear that now the rule is just as above except that, as the last step, we put in a dot so that 6 digits appear to the right of the dot (the actual answer is 7.754702). More generally, if both the decimals have the same number of digits (say 17) after the decimal point, we put a decimal point into the answer so that there are twice as many digits (in our case, 34(= 2 × 17)) to its right.

Finally suppose, for example, that one number has 5 digits to the right of the dot and the other has 2. We add 3 zeros on the right of the second number so that they both have 5 digits to the right of the dot. The resulting product will terminate in 3 zeros and we put in a dot so that 10 digits (including the 3 zeros) are to its right. We throw the 3 zeros away again so that, now, 7 digits are to the right of the dot. You should see that 7 is just 5 + 2. We have our rule!

To multiply two decimals, first forget the decimal points and multiply the resulting whole numbers. Then count the number of digits to the right of the decimal point in each of the two decimals being multiplied and add these two numbers; this is the number of digits to the right of the decimal point in the product.

Example   3.1 × 1.68. First 31 × 168 = 5208. Now 3.1 has 1 digit to the right of the decimal point, 1.68 has 2 digits to the right of the decimal point. 1 + 2 = 3, so we put the decimal point into 5208 so that there are 3 digits to the right of the decimal point. Thus 3.1 × 1.68 = 5.208.

A second approach to the problem, 'What should the product 3.1 × 1.68 be?', is the following. We know that 3.1 is 31 ÷ 10. Thus we feel that 3.1 × 1.68 should be the same as (31 × 1.68) ÷ 10. We know how to multiply 1.68 by the whole number 31, and we know how to divide by 10. Thus we can work out $(31 \times 1.68) \div 10$. You should verify that this gives the same answer, 5.208. However, it is not so natural a way of introducing decimal multiplication because we lose the symmetry (as explained earlier) and because we would have had no reason to suppose that such a product is useful. The product we have defined has been designed to be useful; it enables us to calculate areas (and hence, by repeated multiplication, volumes). Since, however, the two approaches yield the same result, we are now free to use either.

We close this section with a word of warning. That $3.1 \times 1.68 = 5.208$ is perfectly true about decimals, viewed simply as numbers. But if you are told that a rectangular plot of ground is 3.1 m by 1.68 m and you do a calculation and announce that its area is 5.208 m², then you are claiming an accuracy that the measurements do not warrant.† One should certainly say that it is unwise to have more *decimal places* (that is, digits after the dot) in the answer than you had in either of the numbers being multiplied. Thus one should claim no greater accuracy for the area than 5.21 m²; it would be even more modest to leave the result as 5.2 m².

We have seen, then, that decimal calculations are really the same as whole-number calculations; there is just the additional task that we must take care where to insert the decimal point in our numerical answer. All the features of whole number arithmetic are present in decimal arithmetic—including the intense boredom of the actual computations if done by 'hand algorithms'. Fortunately, then, we have the hand calculator available to release us from this boredom and do the tedious calculations for us. The hand calculator can take into account the place of the decimal point in the numbers it processes—but the hand calculator cannot tell us the *meaning* of a decimal calculation nor the *meaning* of the answer. Moreover, even if we know what the calculation means, the hand calculator cannot tell us if the answer it has given us is sensible—we may have made an error in 'punching in' the problem. Such checking of our use of the hand calculator, by means of easy estimates of the sort of answer we should be expecting, is discussed in Chapter 4.

Exercise 1.4

1. Compute the following:

   (a) $.3 \times 6.7 =$ (d) $.0028 \times 1800 =$
   (b) $20 \times .151$ (e) $5.5 \times 1.1 =$
   (c) $1.3 \times 3.1 =$ (f) $20 \times .353 =$

---

†It would in any case be strange to measure the length of a plot of land to a different accuracy from that of the width!

2. **A Puzzle Question** Look at your answers for Problem 1 and make up a problem that would 'fit' for part (g).

3. (a) Mary's rectangular plot of land is 60 m by 40 m. John's plot of land is 84.72 m by 25.28 m. Whose land has the larger perimeter? Whose land has the larger area?

 (b) Make up your own example of such a 'paradox'.

 (c) Suppose you are told that the area of your rectangular plot is 2500 m². What do you think its perimeter is if it is as short as possible? Why?

1.5 PERCENTAGES

We include percentages in a chapter principally devoted to decimals because the ideas are very closely related. There is also a close connection with fractions, but since the fractions that usually enter into percentages are those with a denominator of 100, and since these particular fractions are, as we have explained in Section 1.3, involved in the definition of decimals, it is reasonable to consider the connection between percentages and decimals in this chapter, using these particular fractions merely as intermediaries.

Originally, a percentage was nothing but a fraction of a whole. Thus 100 percent (written 100%) is 100 parts out of 100, or the whole thing. In the same way, 50% is 50 parts out of 100, that is, $\frac{50}{100}$ or $\frac{1}{2}$ of the whole thing. 25% is $\frac{25}{100}$ or $\frac{1}{4}$ of the whole thing, and so on. Now one also uses percentages to indicate fractions bigger than 1. Thus we may be told, of some Latin American country, for example, that it has an inflation rate of 130% per year. What does this mean? It means that an item costing 1 unit price (expressed in that country's currency) one year costs 2.3 units the next year. That

is because 130% means $\frac{130}{100}$, which is 1.3, so that the new price is the old price plus the increase, or 1 + 1.3 = 2.3 units. For example if (heaven forbid!) we had such an inflation rate in the U.S., then gas costing $1 per gallon one year would cost $2.3 per gallon the next year, and bread costing 95¢ a loaf one year would cost (2.3 × 95¢), or $2.19, the next year.

Thus a percentage $P\%$ is just the decimal $P \div 100$. For example, 63% is .63, 212% is 2.12. If we need to compute a percentage of a given amount we multiply together the percentage, expressed as a decimal, and the given amount. The reason we multiply will be discussed at greater length when we make our systematic study of fractions, but an example should make the reason fairly clear. Suppose we wish to calculate 27% of 1436. This means that we are taking $\frac{27}{100}$ of 1436. This means we first take $\frac{1}{100}$ of 1436 and then take 27 of these 'parts'. In other words, we calculate $\frac{1}{100}$ of 1436 and then multiply the result by 27:

$$\frac{27}{100} \text{ of } 1436 \ = \ 27 \times \left(\frac{1}{100} \text{ of } 1436 \right) \ = \ 27 \times (1436 \div 100).$$

From Section 1.3 we know that (1436 ÷ 100) = 14.36, so we could obtain the answer by multiplying 27 × 14.36. Of course, we know from the rule about multiplying decimals in Section 1.4 that we would get *exactly* the same answer if we multiplied .27 × 1436. So,

$$27\% \text{ of } 1436 \ = \ .27 \times 1436 \ = \ 387.72.$$

Having seen the steps above and then observing how they relate to the 'pattern' in this last equation makes it seem quite reasonable (and efficient!) to avoid the steps involving fractions and simply do the computations as follows:

$$143\% \text{ of } 19.2 \ = \ 1.43 \times 19.2 \ = \ 27.456,$$

$$5\% \text{ of } 420 \ = \ .05 \times 420 \ = \ 21.00 \ = \ 21,$$

$$30\% \text{ of } 48.23 \ = \ .30 \times 48.23 \ = \ 14.4690 \ = \ 14.469.$$

Remarks   (i) If an item regularly priced at $45.25 is advertised as being discounted by 25%, it means that the merchant will sell it for $45.25 minus 25% of $45.25. Then we can compute

$$45.25 - (25\% \text{ of } 45.25) \ = \ 45.25 - (.25 \times 45.25)$$
$$= \ 45.25 - 11.3125$$
$$= \ 45.25 - 11.31 \quad \text{(since accuracy to the nearest cent}$$
$$= \ 33.94, \quad \text{is all we need)}$$

and infer that the sale price of the item should be $33.94.

This example illustrates a more general concept that can be stated as follows:

An item regularly selling for *R* dollars, on sale at a discount of *d*%, will have a sale price of

$$R - (d\% \text{ of } R).$$

It is helpful when you read this last expression to

think of '*R*' as the number of dollars for the '*R*egular price,' and *think* of '*d*' as the percentage of the price '*d*iscounted.'

The notation was intentionally chosen to make it easier to remember what the result means. The choice of upper case or lower case letters for '*R*' and '*d*' was arbitrary (although they seemed intuitively suitable, to us) and we could have used '*r*' and '*D*' (or other combinations). We made this particular choice because it did not conflict with any special meanings we already know for the letters involved. Of course, you may use whatever notation is most natural for you.

(ii) Some people would write the calculation in Remark (i) this way:

$$\$45.25 - (25\% \text{ of } \$45.25) = \$45.25 - (.25 \times \$45.25)$$
$$= \$45.25 - \$11.3125$$
$$\text{etc.}$$

This is also correct, but we sometimes prefer to omit the unit signs ('$' is this case) from all of the steps and then simply say that the calculation implies the answer in the units we know to be appropriate. You should remember, however, that if you decide to write the unit signs then they should appear in *all* of the steps. A frequent error in many books is to write the unit signs in some, but not all, of the expressions connected by equal signs. This must surely be a source of anxiety to readers of those books who are paying attention to the meaning of the '=' sign. How, for example, can $3.00 = 3.00? The expression on the left represents an amount of money and the one on the right represents a decimal number!

Exercise 1.5

(Use your hand calculator whenever you feel this will make the problem easier.)
Please perform the following computations: If you do these correctly and examine the results you may see some patterns in the sets of answers.

1. (a) 9% of 55 =
   (b) 170% of 3.5 =
   (c) 50% of 13.9 =

2. (a)   25% of 49.36 =
    (b)   125% of 18.76 =
    (c)   225% of 15.36 =

3. (a)   6% of 18.7 =
    (b)   7% of 31.9 =
    (c)   8% of 41.8 =
    (d)   9% of 49.5 =

4. **A Puzzle Question**   Look at your answers for Problem 3 and make up a problem that would 'fit' for part (e).

5. In parts (a) through (d) an item has a regular price of R dollars and it is on sale at a discount of $d\%$. Find the sale price.

    (a) $R = 99.98, d = 10$.

        (Hint: Complete the following arithmetic, and use the numerical answer to infer the sale price.

        $99.98 - (10\%$ of $99.98) = 99.98 - (.10 \times 99.98) =$ etc.)

    (b) $R = 75.55, d = 25$.
    (c) $R = 96.81, d = 30$.
    (d) $R = 84.82, d = 7$.

6. **A Puzzle Question**   If, because of experience with previous problems in this book, you were looking for a pattern in the answers to Problem 5 you probably felt that something wasn't quite right. Which of the parts doesn't seem to 'fit' in the set of answers? What would have fit?

7. The discount procedure of Remark (i) can also be viewed this way:

        When you receive a discount of 25%, then what you pay is 100% of $45.25 minus 25% of $45.25, hence you pay $(100 - 25)\%$ of $45.25; or, more simply, 75% of $45.25. Then, since

        75% of $45.25 = .75 \times 45.25 = 33.9375,$

    we infer that the sale price, to the nearest cent, should be $33.94.

    (a)   Write the general concept involved here as a statement involving the regular price, R, in dollars, and a discount of $d\%$ (it should look 'like' the boxed statement of Remark (i)).

    (b)   Use the procedure you formalized in part (a) to compute the sale price for each of the parts in Problem 5. (You should, of course, get the same answers!)

2

Fractions

The reader is presumably familiar with fractions as part of our language. Thus we did not hesitate to refer to fractions in Chapter 1 in order to give a workable definition of decimals. In this chapter we discuss the *mathematics* of fractions, and its most important uses. However, we begin with a careful review of how and why fractions are used in our language, since many people feel unsure of some of their uses and meanings.

2.1 REVIEW OF THE NOTION OF A FRACTION

We use words 'a half', 'a quarter', 'a third', 'three-quarters', etc. frequently in everyday speech and their meaning is surely clear to the reader.† Difficulties only arise when we come to do arithmetic with fractions. Let us therefore start from our common knowledge and build on it.

Actually there are subtly different ways in which we use the phrase 'a half'. Offered a piece of cake, we may reply 'Please, just give me a half.' Our hostess or host may then cut the piece of cake into two equal pieces and give us one of those pieces—we have received a half of the original piece of cake. On the other hand, a realtor, showing us two possible lots for purchase may say 'Lot A is more attractively situated but there is only half as much land as on lot B.' There is no suggestion that lot A has been created by cutting up lot B; the realtor only means that the amount of land on lot A is the same as the amount of land we would get by taking half of lot B. When we say that a dime is a tenth of a dollar, we certainly do not mean that a dime is obtained by cutting a dollar bill into ten

---

†We sometimes say 'a fourth' instead of 'a quarter'.

equal pieces and taking one of the pieces; we mean that a dime is worth a tenth of a dollar, that is, we would require 10 dimes to purchase what we can purchase for a dollar.

In mathematics we are really concerned with this broader, and rather more abstract, sense of a fraction. We have in mind some measure and when we say '*A* is three-quarters of *B*' we mean that the measure of *A* is three-quarters of the measure of *B*. The measure may be money ('My income is three-quarters of Bill's'), or length ('My daughter's height is three-quarters of mine'), or weight ('My husband's weight is three-quarters of mine'), or, indeed, any measure at all. Notice that, whereas in the more restricted sense of fraction of which we first spoke ('I've read three-quarters of the book') a fraction is a part of a whole and thus *less* than the whole, in this broader sense it is obviously *not* necessary for the fraction to be less than 1. For example, in our first example, since my income is three-quarters of Bill's, it follows that Bill's income is *greater* than mine, so that if I wish to express Bill's income as a 'fraction' of mine, I must use a fraction greater than 1 (you may already know, or have guessed, that the correct fraction in this case is 'four-thirds' or 'one and a third').

This fact that fractions are used in strictly different senses is frequently overlooked. For example, in one text, fractions are introduced as pieces of a pie and the addition of fractions is referred to the problem of how much pie you have if you take two pieces. One of the examples is $\frac{3}{4} + \frac{3}{8} = \frac{9}{8}$. This is perfectly good arithmetic, but it is nonsense in the context of the pie. Once you have taken three-quarters of a pie, you cannot then take three-eighths *of the same pie*—there isn't enough pie left! We cannot start with a pie and *create* more than we started with by taking impossibly large pieces!

The same confusion is to be found in other texts where, for example we may be told that 'two-thirds of a pie is the same as a third of two pies' and this statement is apparently justified by a nice picture.

Now, in the first, more restricted sense, the statement is evident nonsense! We can take two thirds of a pie without there being another pie in the world! But it *is* true that, if we take two *equal* pies, then the amount of pie we get when we take two-thirds of one of them is the same as the amount of pie we get when we take a third of each of them—the two procedures lead to the same *measure*, but not to the same actual stuff! Since mathematics is concerned with measures, we obviously concentrate on the broader meaning. It is, we hope, plain from what we have said that we must pass from the restricted meaning of fractions as parts of a whole to the broader meaning of fractions as measures of amounts in order to give a meaning to fractions bigger than 1; we will see later that we must do the same if we are to understand equivalent fractions.

However there is a further abstraction we commonly make, and this abstraction brings us face to face with the possibility (we might say, the necessity) of doing some arithmetic with fractions in order to obtain important information. In the phrase '*A* is three-quarters of *B*', *B* may itself be a measure. Thus we may say '*A* is three-quarters of a mile', '*A* is three-quarters of a kilogram', and so on. Thus we may designate the measure of an object by using fractions of a standard measure. Frequently, we are faced with the following type of situation: The price of some commodity is quoted as, say, $48 per unit size. We only require three-quarters of the amount represented by the unit size. How much should we pay? Clearly we should pay three-quarters of $48, but how much is that? This sort of problem becomes even more complicated when we have to take *fractions of fractions*. If the distance round-trip from my home to the super-market is $13\frac{1}{2}$ miles, how far is it one way? The answer is a half of $13\frac{1}{2}$ (in miles), but how far is that? Obviously we need to be able to do some arithmetic with fractions to be able to calculate three-quarters of 48 or a half of $13\frac{1}{2}$.

Before proceeding to answer these questions systematically, let us establish the notation of fractions and the terminology associated with them. We write 'a half' as $\frac{1}{2}$; 'three-quarters' as $\frac{3}{4}$; 'a third' as $\frac{1}{3}$. etc. Now suppose, for example, we want to take

$\frac{2}{3}$ (two thirds) of 60.

We first take an amount, A, which is $\frac{1}{3}$ of 60;
that means 3 times the amount A must equal 60. So $A = 20$.
Then we take 2 times the amount A to get 40.
Thus we see that $\frac{2}{3}$ of 60 is 40.

This example can serve as a 'representative case' of the more general procedure involving any fraction $\frac{N}{D}$, where N and D are whole numbers and D is not zero.† And, in fact, we can use it to write the meaning of the more general statement '$\frac{N}{D}$ of a certain amount B', To do this simply rewrite the above example using

N instead of 2,
D instead of 3,   (so '$\frac{1}{3}$' becomes 'a Dth')
B instead of 60.

Thus the general statement becomes: Suppose we want to take

$\frac{N}{D}$ (N Dths) of a certain number B.

We first take an amount, A, which is 'a Dth' of B;
that means D times the amount A must equal B. So A = ?
Then we take N times the amount A to get C.
Thus we see that $\frac{N}{D}$ of the amount B is C.

Notice that this statement is almost identical in form with the example worked out above. The main difference is that in the generalization we are not able to fill in the values for A and C. The advantage, however, is that this more general statement is true for *all* whole numbers N and D (provided D is not zero). And if we know the values of N, D and B we may be able to find A and C.

Important Remark   We warn you that you may meet, or have met, the notation N/D for the fraction $\frac{N}{D}$. We have not adopted this notation because it has been our experience that students frequently find it confusing, especially when they write mixed numbers. The terms 'top' and 'bottom' (see the footnote) seem much less appropriate when a fraction is written N/D (don't they?). Likewise the rule you will shortly be reading about, in connection with division by a fraction, of 'turning

---

†We call N the *numerator* (or 'top') of the fraction $\frac{N}{D}$ and we call D the *denominator* (or 'bottom') of the fraction. The word 'denominator' is a Latin word which means 'that which names'; the denominator names the size of the parts (in our example, they are thirds). The word 'numerator' is a Latin word which means 'that which counts'; the numerator counts the number of parts (in our example, there are two parts). It is important to notice that a fraction *increases* as we increase the numerator, since we are then increasing the number of parts; but it *decreases* as we increase the denominator since we are then making the parts smaller in size (you get a *bigger* piece of cake if the cake is divided into 3 equal pieces than if it is divided into 4 equal pieces). It would not make any sense to have D equal to zero, since you can't divide any object into zero pieces.

a fraction upside down' seems to require a different description if N and D are written side-by-side. However, if you are comfortable—and careful—with the notation N/D, of course feel free to use it.

Exercise 2.1

Evaluate the following expressions.

1. $\frac{2}{3}$ of 3

2. $\frac{3}{7}$ of 7

3. $\frac{5}{6}$ of 6

4. $\frac{4}{5}$ of 10

5. $\frac{13}{20}$ of 20

6. $\frac{3}{7}$ of 49

7. $\frac{2}{3}$ of 51

8. $\frac{5}{8}$ of 88

9. $\frac{1}{2}$ of 178

10. $\frac{9}{11}$ of 176

11.

12. **For Fun**

 (a)   How can you get the *answer* to Problem 3 by using the *answers* to Problems 1 and 2?

 (b)   How can you get the *answer* to Problem 4 by using the *answers* to Problems 2 and 3?

 (c)   If there had been a Problem 11 what would you guess the answer should be?

2.2 FRACTIONS OF AMOUNTS: MULTIPLICATION OF FRACTIONS

We have already given the example of how we calculate $\frac{2}{3}$ of 60. Suppose, however, that we want to calculate $\frac{2}{3}$ of 40. We meet a difficulty here in deciding what $\frac{1}{3}$ of 40 should be, since there is no whole number X such that if we take 3 times X we get 40. Let us look at an example. What is $\frac{1}{3}$ of 40 yards? Now 40 yards is 120 feet, so $\frac{1}{3}$ of 40 yards is 40 feet. We know that a foot is a third of a yard. We have shown, therefore, that to find $\frac{1}{3}$ of 40 yards, we may first find a third of a yard (a foot) and then take 40 of these measures. But by our definition of a fraction that is precisely $\frac{40}{3}$ of a yard. Now since there is nothing special about the fraction $\frac{1}{3}$, the number 40, or the yard† measure, we can write the following general rule:

$$\frac{1}{D} \text{ of } A \text{ is } \frac{A}{D}$$

†It is special that we happen *already* to have a term 'foot' for the measure which is $\frac{1}{3}$ of a yard. But, for any other measure, nothing prevents us from *inventing* a term for $\frac{1}{3}$ of that measure!

(in our example, $D = 3$, $A = 40$). Notice, however, that in this important rule, something very significant has happened. We expect a fraction of a number to be a number, and yet it has turned out that our answer is a fraction! So we'd better start regarding a fraction as a number! How do we do this? We have noticed that numbers measure amounts and so, in particular, distances. Thus $\frac{1}{3}$ yard is a natural way of writing $\frac{1}{3}$ of a yard and measures a certain distance (1 foot). Thus $\frac{1}{3}$ becomes a number by standing for $\frac{1}{3}$ of the number 1. (This is, of course, a special case of our rule above.) In general, we make a fraction a number by thinking of it as a 'fraction of 1'. This is taken care of automatically in all examples, just as it was in our example that $\frac{1}{3}$ of 40 yards is $\frac{40}{3}$ of a yard. But, if $\frac{40}{3}$ is to be regarded as a number, how big is it? This is, of course, a crucial question, but let us postpone answering it until we have completed our study of the problem of calculating $\frac{2}{3}$ of 40; so far, we've only calculated $\frac{1}{3}$ of 40. In our calculation of $\frac{1}{3}$ of 40 yards we found we could take $\frac{1}{3}$ of a yard and then take 40 of those measures. Thus to calculate $\frac{2}{3}$ of 40 yards, we take $\frac{1}{3}$ of a yard, take 40 of those measures and then take twice what we have. But to take 40 measures, and then take twice that, is to take $2 \times 40 = 80$ measures. So $\frac{2}{3}$ of 40 yards is 80 times the measure of $\frac{1}{3}$ of a yard, or $\frac{80}{3}$ yards. Again, the argument is quite general, so we conclude that we have the rule

Rule 1 $\dfrac{N}{D}$ of A is $\dfrac{N \times A}{D}$

(in our example, $N = 2$, $D = 3$, $A = 40$).

We have made good progress, but we are still in some difficulty over how big a fraction is if it is regarded as a number. Let us look at our earlier example $\frac{40}{3}$. Now the measure $\frac{40}{3}$ yards is the same as 40 feet; and 40 feet is the same as 13 yards 1 foot. This last result is, of course, obtained by dividing 40 by 3, getting a quotient 13 and a remainder 1 (see the Appendix if you are not *absolutely* comfortable with this whole-number arithmetic). Since 1 foot is $\frac{1}{3}$ of a yard, we have shown that $\frac{40}{3}$ yards is the same as $(13 + \frac{1}{3})$ yards, so that

$$\frac{40}{3} = 13 + \frac{1}{3}.$$

The general rule should now be plain. If we have a fraction $\frac{N}{D}$ in which N is bigger than D, and we regard $\frac{N}{D}$ as a number, then we may rewrite it as $Q + \frac{R}{D}$. This is formulated as

Rule 2

$$\frac{N}{D} = Q + \frac{R}{D}$$

Rule for expressing fraction as mixed number

where we divide N by D, getting a quotient Q and a remainder R. Remember that this means that $N = Q \times D + R$.

Notice that, since R is less than D, the fraction $\frac{R}{D}$ is less than 1. As another example we see that

$$\frac{100}{6} = 16 + \frac{4}{6}, \text{ which is also written } 16\frac{4}{6}.$$

This means that if we go $\frac{100}{6}$ of a mile, that is the same as $16\frac{4}{6}$ miles.

We may also 'go the other way'. Suppose we are given the number $Q + \frac{R}{D}$, which we wish to express as a single fraction. We see from Rule 2 than N is just $Q \times D + R$, so that we have

Rule 2′

Rule for expressing mixed number as fraction

$$Q + \frac{R}{D} = \frac{Q \times D + R}{D}.$$

You may feel uncomfortable (and consequently nervous) when you encounter this many unknowns in an expression. This is a natural and normal reaction. Don't panic! What you should do in order to understand what the statement means is to assign numerical values to each of the letters and write out the resulting expression. As an example of applying Rule 2′ let $Q = 7$, $R = 2$ and $D = 3$. Then we see that

$$7 + \frac{2}{3} = \frac{7 \times 3 + 2}{3} = \frac{21 + 2}{3} = \frac{23}{3}.$$

You may wish to select other numbers for Q, R and D. Do this until you feel comfortable about the meaning of Rule 2′. More importantly, remember to use this technique in the rest of the book whenever there are unknowns in the expressions and you feel unsure about what the expression really means.

Many texts use the terminology 'proper fraction' for a fraction $\frac{N}{D}$ in which N is less than D, 'improper fraction' for a fraction in which N is greater than (or equal to) D, and 'mixed number' for the expression on the right hand side of Rule 2. In our view, the main advantage of such terms is that it makes it easier for the textbook writer to set the reader problems! They also help to explain rules like Rule 2 and Rule 2′ and other aspects of the arithmetic of

fractions. Mathematicians use a very different terminology (fractions and rational numbers); this will be described in the next section.

So far we have seen how to obtain any fraction of a whole number, and how to find out how big a fraction is when regarded as a number. Our final objective in this section is to show how to obtain a fraction of a fraction. We will go through the argument very carefully, but some of you may be content to just look at the illustration.

Suppose we want to find $\frac{3}{4}$ of $\frac{7}{10}$. Let us take as our concrete example $\frac{3}{4}$ of $\frac{7}{10}$ of a meter. Now $\frac{7}{10}$ of a meter is 7 decimeters (since a decimeter is $\frac{1}{10}$ of a meter), so we must calculate $\frac{3}{4}$ of 7 decimeters and Rule 1 tells us this is $\frac{21}{4}$ decimeters. How many meters is this? To answer this question let us suppose we have a stick whose length is $\frac{1}{4}$ of a decimeter. Then 4 sticks have a combined length of 1 decimeter, so 40 sticks have a combined length of 1 meter; and $\frac{21}{4}$ decimeters is the length of 21 sticks. But 1 stick has length $\frac{1}{40}$ of a meter, so 21 sticks have length $\frac{21}{40}$ of a meter. Thus $\frac{3}{4}$ of $\frac{7}{10}$ of a meter is $\frac{21}{40}$ of a meter, or

$$\frac{3}{4} \text{ of } \frac{7}{10} \text{ is } \frac{21}{40}.$$

How did we get this? If we look carefully at the argument we see that the '21' in the numerator came (by Rule 1) from multiplying 3×7 and the '40' in the denominator came from multiplying 4×10. Thus our general rule is

Rule 3 $\frac{N}{D}$ of $\frac{A}{B}$ is $\frac{N \times A}{D \times B}$.

We have shown that to take a fraction of a fraction, we must multiply the numerators to get the numerator of the resulting fraction, and multiply the denominators to get the denominator of the resulting fraction. For example,

$$\frac{1}{2} \text{ of } \frac{3}{4} \text{ is } \frac{3}{8},$$

since $1 \times 3 = 3$, $2 \times 4 = 8$. It seems reasonable to *call* the process, of multiplying numerators to get a numerator and multiplying denominators to get a denominator, *the multiplication of fractions.* It is *very important* to understand that we are now inventing a meaning for the multiplication of fractions—nothing in our previous arithmetic could have told us how to multiply fractions. We have chosen the meaning so that it is (a) *natural,*

Definition of multiplication

$$\frac{N}{D} \times \frac{A}{B} = \frac{N \times A}{D \times B}$$

and (b) *useful,* since we can now use it to write out a fraction of a fraction,

Rule 3
(restated)

$$\frac{N}{D} \text{ of } \frac{A}{B} \text{ is } \frac{N}{D} \times \frac{A}{B}.$$

A particularly important special case of Rule 3 is

$$\frac{1}{D} \text{ of } \frac{A}{B} \text{ is } \frac{A}{D \times B}.$$

Now to take $\frac{1}{D}$ of a number you divide that number by D. Thus, to *divide a fraction by D we multiply its denominator by D.* Notice that we have already shown that $\frac{1}{D}$ of A is $\frac{A}{D}$, so that the fraction $\frac{A}{D}$ (viewed as a number) may be thought of as obtained by dividing A by D. This is consistent with what we have just said since A is the same as $\frac{A}{1}$. (Of course, we don't often actually *write* $\frac{A}{1}$, but it is important to notice that a whole number can be regarded, in this way, as a fraction, just as, in the previous chapter, we remarked that a whole number can be regarded as a decimal.)

Exercise 2.2

In Problems 1 through 6 use Rules 1 and 2 to determine how big the expression is as a number. (Resist the temptation to 'reduce' your answers, if you know how to do that—and don't worry if you don't know what that means.)

1. $\frac{3}{8}$ of 17

2. $\frac{3}{10}$ of 28

3. $\frac{5}{12}$ of 25

4. $\frac{6}{14}$ of 29

5. $\frac{7}{16}$ of 33

6. $\frac{8}{18}$ of 37

7.

8. **A Puzzle Question**

    (a)   Write a replacement for Problem 2 that will fit better in the sequence of Problems 1 through 6.

    (b)   Make up an appropriate Problem 7 for this sequence, and check to see if the answer fits the sequence of answers.

In problems 9 through 14 apply the 'reverse' Rule 2' to write the given expressions as a single fraction. If necessary continue to *pretend* you don't know how to 'reduce' fractions (that is covered in the next section).

9. $5 + \dfrac{1}{2}$ (or $5\dfrac{1}{2}$)   12. $3\dfrac{8}{11}$

10. $4 + \dfrac{1}{5}$ (or $4\dfrac{1}{5}$)   13. $3\dfrac{9}{14}$

11. $3\dfrac{7}{8}$ 14. $3\dfrac{10}{17}$

15.

16.   **For Fun**

    (a)   Look at the numerators in Problems 9 through 14 and guess what the numerator in the answer to Problem 15 should be.

    (b)   Look at the denominators in Problems 9 through 14 and guess the denominator that would 'fit' for Problem 15.

    (c)   Rewrite Problems 9 and 10 so that they fit this problem sequence better.

    (d)   Make up a Problem 15 so that it fits the sequence and check, using Rule 2', to see if it fits your guesses for parts (a) and (b).

In Problems 17 through 20 first use Rule 3 to multiply the fractions; then use Rule 2 to find out how big that fraction is as a number.

17. $\dfrac{43}{4} \times \dfrac{4}{15} =$

18. $\dfrac{9}{2} \times \dfrac{8}{30} =$

19. $\dfrac{61}{20} \times \dfrac{2}{3} =$

20. $\dfrac{6}{5} \times \dfrac{37}{12} =$

21.   (a)   Write the final answers in Problems 17 through 20 in a sequence so that each number is smaller than the next. Do you see a pattern?

    (b)   Write the fractional form of the answers in Problems 17 through 20 in a sequence so that each number is smaller than the next. Now do you see a pattern?

2.3 EQUIVALENT FRACTIONS

Let us return to a calculation we did in the previous section. Suppose we have a 'stick measure', so that

4 sticks = 1 decimeter.

We know that

10 decimeters = 1 meter.

Then

12 sticks = 3 decimeters = $\frac{3}{10}$ meter.

But it is also true that

40 sticks = 1 meter,

so

1 stick = $\frac{1}{40}$ meter

and

12 sticks = $\frac{12}{40}$ meter

Thus we see that

$$\frac{3}{10} = \frac{12}{40}.$$

We could also see this in a different way. What does $\frac{4}{4}$ mean? It means that we divide an object into 4 equal parts and take 4 of those parts—we have the *whole* object. Thus $\frac{4}{4}$, as a number, is just the number 1. Then

$$\frac{3}{10} \text{ of } \frac{4}{4} \text{ is } \frac{3}{10} \text{ of 1, or } \frac{3}{10}.$$

But, by Rule 3,

$$\frac{3}{10} \text{ of } \frac{4}{4} \text{ is also } \frac{3 \times 4}{10 \times 4} \text{ so } \frac{3 \times 4}{10 \times 4} = \frac{3}{10}.$$

You will see, especially from this last piece of reasoning, that we have again obtained a result of great generality.

Rule 4 $\qquad \dfrac{N \times K}{D \times K} = \dfrac{N}{D}$ (where $K \neq 0$).

Should we say that

$$\frac{N \times K}{D \times K} \text{ and } \frac{N}{D}$$

are the *same* fraction? Are $\frac{12}{40}$ and $\frac{3}{10}$ the same fraction? Strictly speaking the answer must be 'no'. For fractions have numerators and denominators (tops and bottoms) and clearly $\frac{12}{40}$ and $\frac{3}{10}$ have different tops and different bottoms. *So the fractions $\frac{12}{40}$ and $\frac{3}{10}$ are different but represent the same number* (and the same proportion of a given object). The analogy here is with names. Imagine the following telephone conversation: 'Is that Mr. Brown?' 'Yes, this is Mr. Brown speaking.' 'Hi, Jim, this is Paul.' What can we infer? Paul wanted to talk to 'Mr. Brown' and then addressed him as 'Jim', another name for the same person. In just the same way $\frac{12}{40}$ and $\frac{3}{10}$ are different names for the same number. A fraction is thus a name of a number (or of a measure or proportion). This is not special to fractions; after all $\frac{3}{10}$ also has the names .3, 0.3. We won't keep stressing this nominal aspect of fractions, but it will be useful to be able to refer to those numbers that can be named by fractions. Remember that our original set of numbers 0, 1, 2, 3, · · · is called the set of *whole* numbers and we have now enlarged our set of numbers. Numbers named by fractions or mixed numbers are called *rational*† numbers. Since, as already pointed out, $4 = \frac{4}{1}$, we see that every whole number is a rational number, but, of course, not every rational number is a whole number—otherwise we would have had no need to introduce fractions in the first place! To indicate that two fractions represent the same rational number, we call them *equivalent fractions*. Rule 4 tells us that, given any fraction, we obtain an equivalent fraction by multiplying the top and bottom by the same non-zero whole number. It also tells us that we get an equivalent fraction by dividing top and bottom by the same non-zero whole number (provided both top and bottom are exactly divisible by that whole number—otherwise we don't get a fraction at all). We will shortly produce a rule which tells us precisely when two fractions are equivalent.

FRACTION   REDUCED
             FRACTION

Passing from $\frac{12}{40}$ to $\frac{3}{10}$ is called *reducing* the fraction. It often simplifies a calculation (especially if it is sufficient to do the calculation approximately) to reduce the fraction. Thus if we want $\frac{12}{40}$ of 300 it is much easier to rewrite this as $\frac{3}{10}$ of 300, which is quickly seen to be 90. However reducing a fraction is not *always* a good idea. If we want $\frac{26}{100}$ of 213, it is best simply to multiply 26 × 213, getting 5538 and then 'put in the dot' in the appropriate place, getting 55.38. It would be distinctly harder first to reduce the fraction to $\frac{13}{50}$ and then work out $\frac{13}{50}$ of 213.

Of course, the reduction may well take place at a later stage of the calculation. If we want

$$\frac{11}{14} \text{ of } \frac{91}{22}$$

†The word 'rational' is based on the notion of ratio. It has nothing to do here with 'rational' in the sense of 'reasonable'.

$$\frac{423}{718}$$

Rather than wrestle with large numbers, seek the help of a hand calculator.

we must simply use Rule 3 since neither fraction can be reduced, as you may yourselves verify. Thus

$$\frac{11}{14} \text{ of } \frac{91}{22} \text{ is } \frac{11 \times 91}{14 \times 22}.$$

Now, however, we have a fraction which certainly can be reduced. We can divide the top and bottom by 11, getting

$$\frac{91}{14 \times 2},$$

and then divide top and bottom by 7, getting

$$\frac{13}{2 \times 2} \quad \text{or} \quad \frac{13}{4}.$$

This then is the answer in the simplest fraction form (another 'simple' form is $3\frac{1}{4}$, another is 3.25, another is 325%). When we can reduce a fraction no further we say it is *completely reduced* or, in more old-fashioned language, *in its lowest terms*. We handle a reduction usually by a process of 'crossing out'. Thus, typically, the calculation given above would be displayed as

$$\frac{\overset{1}{\cancel{11}} \times \overset{13}{\cancel{91}}}{\underset{2}{\cancel{14}} \times \underset{2}{\cancel{22}}} = \frac{13}{4}$$

How can we spot the existence of a common factor of the numerator and denominator of a fraction, so that we may reduce the fraction? The ability to do this comes with experience and familiarity with numbers. There are systematic procedures but they tend to be tedious. The best advice is probably to try to spot such common factors and, if you fail, use a hand calculator to do the calculation if it looks complicated.

We now take up the question of deciding exactly when two fractions are equivalent. Notice that two fractions may be equivalent and yet neither can be gotten by reducing the other—a simple example of this is the pair of fractions $\frac{4}{6}$ and $\frac{6}{9}$. If we look at this example we recognize they are equivalent because each reduces to $\frac{2}{3}$, so we might conjecture that two fractions are equivalent if (and, of course, only if) they both reduce to the same fraction. This will turn out to be true, but it is still not entirely satisfactory because, as we have said above, it may be difficult and tedious to check whether the two fractions do indeed reduce to the same fraction. We will now develop a more effective test.

We first note that if two fractions $\frac{M}{D}$ and $\frac{N}{D}$, with the same denominators, are to be equivalent, they must have the same numerators (that is, $M = N$). For if we divide an object into D equal parts, we can only get the same amount when we take M of those parts as when we take N of those parts if $M = N$. We use this remark to get ourselves a general rule for deciding when two fractions are equivalent. Let's look at the fractions

$$\frac{4}{6} \text{ and } \frac{6}{9}.$$

Then

$$\frac{4}{6} \text{ is equivalent to } \frac{4 \times 9}{6 \times 9} = \frac{36}{54},$$

and

$$\frac{6}{9} \text{ is equivalent to } \frac{6 \times 6}{9 \times 6} = \frac{36}{54}.$$

Thus

$$\frac{4}{6} \text{ and } \frac{6}{9} \text{ are equivalent fractions.}$$

Again, if we take $\frac{5}{6}$ and $\frac{7}{9}$ then we would find that $\frac{5}{6}$ is equivalent to $\frac{45}{54}$ and $\frac{7}{9}$ is equivalent to $\frac{42}{54}$, so $\frac{5}{6}$ cannot be equivalent to $\frac{7}{9}$ because, obviously, $\frac{45}{54}$ is not equivalent to $\frac{42}{54}$, the numerators being different while the denominators are the same.

This type of argument, applied to any pair of fractions, leads to the following general rule.

Rule 5   The fractions $\frac{A}{B}$ and $\frac{C}{D}$ are equivalent if, and only if,

$A \times D = B \times C$. For example, if we apply Rule 5 to $\frac{4}{6}$ and $\frac{6}{9}$

we find that $4 \times 9 = 6 \times 6 \ (= 36)$, so that $\frac{4}{6}$ and $\frac{6}{9}$ are

equivalent.

* We close this section by giving the argument which shows that, if two fractions are equivalent, they must each reduce to the same fraction. You may ignore this proof if you prefer to pass straight to the next section. Suppose then that $\frac{A}{B}$ reduces to $\frac{X}{Y}$ and that $\frac{C}{D}$ reduces to $\frac{U}{V}$, and that $\frac{X}{Y}$ and $\frac{U}{V}$ are completely reduced. Then $\frac{X}{Y}$ and $\frac{U}{V}$ are equivalent if $\frac{A}{B}$ and $\frac{C}{D}$ are equivalent, so that $X \times V = Y \times U$. This means that X is a factor of $Y \times U$. But X and Y have no factors in common, so that X is, in fact, a factor of U. But U is a factor of $X \times V$ and U and V have no factors in common so that U is a factor of X. So X and U are two whole numbers with each a factor of the other. This forces them to be equal, $X = U$. Similarly, $Y = V$, so that the fractions $\frac{X}{Y}$ and $\frac{U}{V}$ are the same. You should notice that we have shown in the course of this argument that reducing a fraction $\frac{A}{B}$ so that it becomes completely reduced leads to a *unique* completely reduced fraction—not a surprising result, but mathematicians like to have proofs even when the conclusion seems obvious.

Exercise 2.3

1. Determine which of the following fractions are equivalent to each other.

(a) $\frac{2}{3}$ (g) $\frac{4}{9}$

(b) $\frac{1}{2}$ (h) $\frac{107}{214}$

(c) $\frac{5}{7}$ (i) $\frac{35}{49}$

(d) $\frac{8}{12}$ (j) $\frac{12}{27}$

(e) $\frac{22}{33}$ (k) $\frac{34}{51}$

(f) $\frac{45}{63}$ (l) $\frac{48}{108}$

2. Carry out the following multiplications. Try to reduce your answers completely; but notice that you can still check to see if your answer is equivalent to the answer given (and therefore perfectly correct) even if it is not completely reduced. You can use the given answer to find the

numbers that will divide into the top and bottom numbers of your answer.

(a) $\dfrac{12}{15} \times \dfrac{5}{6} =$ (d) $\dfrac{4}{6} \times \dfrac{9}{10} =$

(b) $\dfrac{6}{77} \times \dfrac{55}{6} =$ (e) $\dfrac{202}{1717} \times \dfrac{39}{6} =$

(c) $\dfrac{98}{77} \times \dfrac{15}{30} =$ (f) $\dfrac{10}{65} \times \dfrac{649}{118}$

3. (a) Arrange the completely reduced answers for Problem 2 in a sequence so that the top number in each fraction is less than the top number in the next fraction.

(b) Can you write any of the numbers in either the top or the bottom of these fractions as the product of two whole numbers, where neither of the numbers are 1?

(c) Starting with 2, and ending with 29, write in ascending order all of the whole numbers that *cannot* be written as the product of two whole numbers unless you use the number 1 as a factor. (For example, the only way you can write 3 as the product of two whole numbers is as either 3×1 or 1×3. But, if you are not allowed to use 1 then it cannot be done. Thus 3 satisfies our requirement. But $4 = 2 \times 2$, so 4 does not fit the requirements.)   Numbers having this property are called *prime* numbers.

(d) What would be the next fraction in the sequence you wrote for part (a)?

4. When you want to reduce fractions it is helpful to recognize when a number is divisible by some of the first prime numbers like 2, 3 and 5. For example if a whole number ends with a 0 or a 5, then 5 will divide into that number with no remainder.

(a) How do you tell if a number is divisible by 2?

(b) Here are some numbers that are divisible by 3.

111, 1101, 1011, 100002, 105, 204, 27, 108, 3009, 1836, 5514

Find the sum of the digits for each number (for 111, you get $1 + 1 + 1 = 3$) and make a guess about how you can tell if a number is divisible by 3. Then try some numbers of your own choosing and *check your guess.*

(c) Here are some numbers that are divisible by 9.

27, 108, 7128, 3438, 876312, 88911

Find the sum of the digits for each number and make a guess as to how you can tell if a number is divisible by 9. Select some numbers of your own and see if your guess works.

5. Write the following fractions in completely reduced form.

(a) $\dfrac{2349}{621}$ (c) $\dfrac{45927}{72171}$

(b) $\dfrac{60}{630}$ (d) $\dfrac{1260}{8580}$

2.4 FRACTIONS AND DECIMALS

Many texts assert 'Decimals and fractions are the same things, but written differently.' This is untrue. Every decimal can, as we saw in Chapter 1, be written as a fraction; but the only fractions which can be written as decimals are those equivalent to fractions whose denominators are powers of 10. Thus, for example, $\frac{1}{4}$ may be written as a decimal, since $\frac{1}{4}$ is equivalent to $\frac{25}{100}$; but $\frac{1}{3}$ cannot be written as a decimal. Nevertheless these same texts may say that $\frac{1}{3}$ can be written as the *infinite decimal*

$$\frac{1}{3} = 0.333\cdots$$

What does this mean? Certainly an 'infinite decimal' is not a decimal in the sense of Chapter 1, since our rules for adding, subtracting and multiplying decimals do not apply to infinite decimals; it is essential to our whole concept of a decimal that it is *not* 'infinite' so that we can regard decimals as arising in real life by adopting different units of measurement (for example, 27 meters = 0.027 kilometers in the metric measure of distance). The mysterious assertion $\frac{1}{3} = 0.333\cdots$ means that we can get as near to $\frac{1}{3}$ as we like by taking a decimal with a sufficient number of 3's after the decimal point. Thus

$$\frac{1}{3} - 0.3 = \frac{1}{30},$$

$$\frac{1}{3} - 0.33 = \frac{1}{300},$$

$$\frac{1}{3} - 0.333 = \frac{1}{3000}, \text{ etc.}†$$

Here we introduce a key idea in mathematics: *approximating to a rational number by a decimal.* The hand calculator and the computer use this principle. Ask a hand calculator to calculate 5 ÷ 3 and it will produce a display of 1 followed by the decimal point, followed by a whole string of 6's, followed by 7. The '7' arises since this gives the best possible approximation to $\frac{5}{3}$ to the accuracy of which the hand calculator is capable. Thus 1.67 is closer to $\frac{5}{3}$ than 1.66 because $\frac{5}{3}$ is bigger than 1.666. So the truth of the matter is that every decimal may be rewritten as a fraction, and every fraction may be approximated by a decimal to whatever accuracy is needed. We will say no more about infinite decimals in this text (except for a brief reference in Problem 6 at the end of this section).

When we consider the *utility* of fractions and decimals, we find that they really play very different roles in our quantitative thinking.

---

†This kind of computation will be discussed in Section 2.5. It is shown here simply to illustrate the *meaning* of 0.333 ⋯.

Decimals typically arise in measurement (including 'measures' of money) and, since we so often find ourselves adding and subtracting measurements, it is natural that the addition and subtraction of decimals is of great importance (and easy!). Certain measurements do get multiplied and divided so we do also have to multiply and divide decimals; but such matters arise less frequently and are more sophisticated.†

Fractions on the other hand typically arise in questions of taking a certain number of parts of a whole, that is, in matters of ratio and proportion. Thus, in every-day speech a fraction is almost always followed by the word 'of,' whereas a decimal (like $5.27) practically never is. Since taking fractions 'of something' involves us, as we saw in Section 2.2, in multiplication, we find we often have to multiply fractions. But we rarely add or subtract proportions, so adding and subtracting fractions is less important†† than multiplying them (and much harder!). So decimals and fractions play different roles and this is reflected in a difference between the most important aspects of their arithmetic. This distinction means that mathematics would depart too far from real life if it insisted that we regard fractions and decimals as essentially the same thing. Put another way, we *can* replace decimals by fractions but it is not often the case that we benefit by doing so.

Textbooks often contain phony problems, in a blatant attempt to justify the presence in the standard curriculum of some piece of arithmetic which really has, at best, very limited utility. The following example is taken from a much-used text. 'Johnny swam $43\frac{2}{7}$ meters on Monday and $38\frac{1}{5}$ meters on Tuesday. How far did he swim altogether on the two days?' The problem is, of course, supposed to 'justify' the time spent teaching the rules for adding fractions; but it is absolutely phony. Neither Johnny nor anybody else could possibly be in possession of the astonishing facts quoted in the problem! What swimming pool is calibrated in fifths and sevenths of meters!? No, the problem *should* be one of adding decimals, but the authors of the textbook feel themselves under an obligation to give an 'application' of the addition of fractions and so give the information about Johnny's swimming achievements in an absurd, unnatural and entirely artificial form.

However, since decimals *may* be converted into fractions, and since we have given a rule for multiplying decimals (Section 1.4) and a rule for multiplying fractions (Section 2.2), we are obliged to show that the rules are consistent. One way to do this—this will be the way

---

†We deal with division in the next chapter.

††We do need to add fractions in studying the theory of *probability*. This will be taken up in Section 2.5.

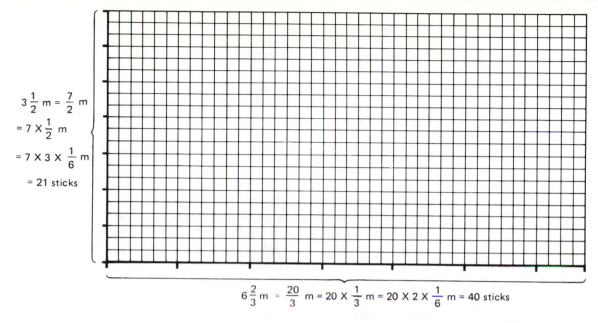

$3\frac{1}{2}$ m $= \frac{7}{2}$ m

$= 7 \times \frac{1}{2}$ m

$= 7 \times 3 \times \frac{1}{6}$ m

$= 21$ sticks

$6\frac{2}{3}$ m $= \frac{20}{3}$ m $= 20 \times \frac{1}{3}$ m $= 20 \times 2 \times \frac{1}{6}$ m $= 40$ sticks

we choose—is to show that if we *do* measure distances in, say, fractions of a meter, then areas of rectangles may be calculated in square meters by multiplying the fractions by the rule given in Section 2.2. Let us now show this to be true, but the reader should understand that what follows is designed to show the consistency of our rules for multiplying fractions and decimals, and *not* to give a genuine application of the multiplication of fractions.

Suppose then that a rectangular plot is $3\frac{1}{2}$ meters long and $6\frac{2}{3}$ meters wide and we wish to calculate its area. We must first express its length and width as fractions (of a meter). By Rule 2' we see that

$$3\frac{1}{2} = \frac{3 \times 2 + 1}{2} = \frac{7}{2}, \quad \text{and} \quad 6\frac{2}{3} = \frac{6 \times 3 + 2}{3} = \frac{20}{3}.$$

Thus the length of the plot is $\frac{7}{2}$ m, the width is $\frac{20}{3}$ m, and we want to show that its area is $(\frac{7}{2} \times \frac{20}{3})$ m² (square meters) or $(\frac{7 \times 20}{6})$ m², according to our definition of the multiplication of fractions. Let us invent a new unit of length, the stick, so that 6 sticks = 1 meter. (We chose 6 because $6 = 2 \times 3$, the product of the two denominators in our fractions). Then $\frac{1}{2}$ m = 3 sticks, so that $\frac{7}{2}$ m = 7×3 sticks; and $\frac{1}{3}$ m = 2 sticks, so that $\frac{20}{3}$ m = 20×2 sticks. Thus the area of the plot is $(7 \times 3 \times 20 \times 2)$ square sticks. But since there are 6 sticks to a meter, there are (6×6) square sticks to a square meter. So

1 m

1 m

$$(7 \times 3 \times 20 \times 2) \text{ square sticks } = \frac{7 \times 3 \times 20 \times 2}{6 \times 6} \text{ m}^2$$

$$= \frac{7 \times 20}{6} \text{ m}^2, \text{ using Rule 4.}$$

Once again, since there was nothing special about the numbers used in this computation, our argument is really completely general, so that the argument does establish the consistency of decimal and fraction multiplication.

Since only fractions with denominators that are powers of 10 actually arise in converting decimals to fractions, you might have expected us to confine attention in these last remarks to such fractions. The reason we do not is that, by our more general approach involving any fraction whatsoever, we establish an important principle of *approximation*, namely, if we are required to multiply two fractions, we may approximate to the fractions by decimals, multiply the decimals, and obtain in this way an approximation to the product of the fractions. This can be a useful strategy if we must do the calculations by hand, though it would not be worthwhile adopting it *explicitly* if we have a hand calculator available. However, it is essentially what most hand-calculators actually do in multiplying fractions.

However, you may feel, especially in the light of what we have said about decimals and fractions each having their proper place in our mathematical language, that it would be unnatural to present a decimal as the product of two fractions. This objection is valid if an *exact* answer is required; but it is typical of decimals that they arise when we are content to give approximate answers.†

There is one type of mathematical expression in which fractions and decimals naturally occur together. Since we may take fractions of anything measurable, we may of course take fractions of decimals. Thus such a phrase as 'a half of 12.4' arises often, in connection with length, weight and other measures. To calculate such a quantity, we could, of course, convert the decimal to a fraction and proceed by means of Rule 3. However, this would not be natural, since it would yield a fraction as our answer and we are expecting a decimal. Thus the correct procedure, in view of the real-life context in which the problem arose, would involve us in dividing a decimal by a whole number, a subject taken up in the next chapter. Of course, in our particular example, there is no problem—the answer is clearly 6.2. However if we required a third of 12.4 instead of a half, there would be a subtle question involved, closely related to our discussion of approximating fractions by decimals.

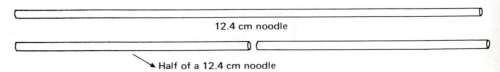

12.4 cm noodle

Half of a 12.4 cm noodle

---

†It is not, however, always the case. It would be natural to approximate the fraction $\frac{101}{300}$, in certain contexts, by the fraction $\frac{1}{3}$.

Exercise 2.4

1. Look at an elementary textbook and see how many 'phony' or absurd problems you can find involving the addition of fractions. Also note the really *useful*, or motivating, examples and problems that involve the addition of fractions.

2. Find the result of multiplying 1.2 by .9
   (a) by multiplying the numbers as *decimals*;
   (b) by converting the numbers to fractions, multiplying the fractions, and converting the resulting answer to a mixed number;
   (c) by drawing an accurate picture of a rectangle 1.2 units by .9 units and counting the number of square units. (You may wish to calibrate the units along the sides in tenths.)

3. Find the following products by the method of your choice. Write your answer in the form that seems the most natural to you.
   (a) $.41 \times .3 =$
   (b) $\dfrac{3}{29} \times \dfrac{4}{3} =$
   (c) $\dfrac{2}{13} \times 3\dfrac{2}{5} =$
   (d) $.6 \times .39 =$
   (e) $15 \times .023 =$
   (f) $8 \times \dfrac{7}{43} =$
   (g) $.24 \times 1.9 =$

4. **Pure Fun**
   (a) Write in a sequence the first form of the answers *given* in the back of the book for parts (a), (d), (e) and (g) of Problem 3.
   (b) Guess what the next number in this sequence should be. (Notice how obscure this pattern would have been if you had written this sequence using the last answer given in each case.)

5. **More Fun**
   (a) Write in a sequence the first form of the answers given in the bakc of the book for parts (b), (c) and (f) of Problem 3.
   (b) What would be the next fraction in this sequence?

6. The instructions for some hand calculators have a statement similar to the following:

If the display shows	the answer is
.99999999	1
1.9999999	2
2.9999999	3
.	.
.	.
.	.

The expressions on the left are really infinite decimals, as described in the beginning of the section, but the display only has room for eight digits. Complete the following decimal computations to see why this is a reasonable way to interpret the display.

(a) $1 - .9 = 1.0 - .9 = .1$
      $1 - .99 = 1.00 - .99 =$
      $1 - .999 =$
      $1 - .9999 =$

(b) $2 - 1.9 = 2.0 - 1.9 = .1$
      $2 - 1.99 =$
      $2 - 1.999 =$
      $2 - 1.9999 =$

A Hand Calculator Challenge†

In problems 7, 8 and 9 perform the computations with a hand calculator. Write your answers in an orderly way.

7. (a) $\frac{1}{3} =$ and $\frac{1}{33} =$

    (b) $\frac{1}{9} =$ and $\frac{1}{11} =$

8. (a) $\frac{1}{3} =$ and $\frac{1}{333} =$

    (b) $\frac{1}{9} =$ and $\frac{1}{111} =$

    (c) $\frac{1}{27} =$ and $\frac{1}{37} =$

9. (a) $\frac{1}{3} =$ and $\frac{1}{3333} =$

    (b) $\frac{1}{9} =$ and $\frac{1}{1111} =$

    (c) $\frac{1}{11} =$ and $\frac{1}{909} =$

    (d) $\frac{1}{33} =$ and $\frac{1}{303} =$

    (e) $\frac{1}{99} =$ and $\frac{1}{101} =$

10.

11. Make up the appropriate Problem 10. (Hint #1: There are 5 parts to the problem) (Hint #2:

      Problem 7: $3 \times 33 = ?$
                  $9 \times 11 = ?$
      Problem 8: $3 \times 333 = ?$
                  $9 \times 111 = ?$
                  $27 \times 37 = ?$
      Problem 9: $3 \times 3333 = ?$
             etc.   )

   (Hint #3: $41 \times 271 = 11111$.)

†The idea for this set of problems came from Problem 15.45 of G. Pólya's book, *Mathematical Discovery* (combined edition), *Vol II* (p. 165), John Wiley & Sons, 1981.

2.5 ADDITION OF FRACTIONS

So far we have avoided discussing the addition (and subtraction) of fractions, but we can no longer postpone treating this subject, which is so often the despair of the student. When do we need to add fractions? There are two situations in which this arises, of which one is far more important from the point of view of the utility of arithmetic. However, we will first discuss the *less* useful application! The more important application, to probability theory, will be presented in the next section.

It may happen that we do record measurements using fractions, so that the addition of measurements can give rise to the addition of fractions. We have already discussed in this chapter the possibility of using fractions in, say, measuring lengths, and we have pointed out that it is very often absurd to do so. Unfortunately we continue, in this country, to use the American system of weights and measures a great deal, and, in this cumbersome and inefficient system, we may find measures of length and weight recorded in fractional measures. The fractions we would tend to use in this connection would most frequently be the simplest fractions ($\frac{1}{2}$, $\frac{1}{4}$, $\frac{3}{4}$, $\frac{1}{3}$, $\frac{2}{3}$), so that we would only have to add such fractions in most applications in this category. Let us give an example, *not chosen for its simplicity,* to obtain a general rule for adding fractions.

Suppose the distance from A to B is $28\frac{1}{2}$ miles and the distance from B to C is $34\frac{2}{5}$ miles. What is the distance from A to C, via B? We do not suggest that this is a 'good' application, since we might, say, know the distances from the odometer on our car and they would have then been given to us as 28.5 miles and 34.4 miles. We would then add the decimals to get the answer 62.9 miles. However, our example is at least easily understood and so can serve as an *illustration* of the addition of fractions. The method we will describe could be used when the fractions do not have exact decimal equivalents.

We first remark that the total distance is obviously $(28 + 34)$ miles plus $(\frac{1}{2} + \frac{2}{5})$ miles, so that our problem is to add $\frac{1}{2} + \frac{2}{5}$. Once

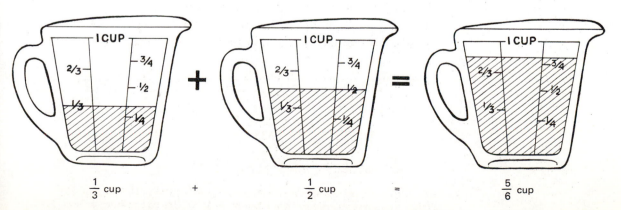

$$\frac{1}{3} \text{ cup} \qquad + \qquad \frac{1}{2} \text{ cup} \qquad = \qquad \frac{5}{6} \text{ cup}$$

we can do that, we only have to add the result to 62 miles. The point we are making is that our *new* arithmetic problem is that of the addition of fractions less than 1 (*proper fractions,* as they are sometimes called). In the problem we are considering, the distances are already given in the form of a *mixed number* (whole number + proper fraction), which is convenient.

So we must add $\frac{1}{2} + \frac{2}{5}$. We invent a new measure of distance, the stick, given by 10 sticks = 1 mile; we chose this measure because $10 = 2 \times 5$, where 2 and 5 are the denominators of the fractions being added. Then $\frac{1}{2}$ mile = 5 sticks, $\frac{2}{5}$ mile = 4 sticks, so that $\frac{1}{2}$ mile $+ \frac{2}{5}$ mile = 9 sticks. Since 1 stick = $\frac{1}{10}$ mile, 9 sticks = $\frac{9}{10}$ mile. Thus $\frac{1}{2} + \frac{2}{5} = \frac{9}{10}$, and the answer to our problem is that the distance from A to C, via B is $62\frac{9}{10}$ miles; this, of course, agrees with the answer we obtained using the addition of decimals.

Let us now analyze the method we used to show that $\frac{1}{2} + \frac{2}{5} = \frac{9}{10}$. We invented the stick, which is $\frac{1}{10}$ of a mile and said that $\frac{1}{2}$ mile = 5 sticks, $\frac{2}{5}$ mile = 4 sticks. That is,

$$\frac{1}{2} \text{ mile} = \frac{5}{10} \text{ mile}, \qquad \frac{2}{5} \text{ mile} = \frac{4}{10} \text{ mile},$$

or

$$\frac{1}{2} = \frac{5}{10}, \qquad \frac{2}{5} = \frac{4}{10},$$

You should observe that what we have effectively done is to replace the fractions $\frac{1}{2}$, $\frac{2}{5}$ by equivalent fractions *having the same denominator,* that is, $\frac{5}{10}$, $\frac{4}{10}$ respectively. Then it is plain that $\frac{5}{10} + \frac{4}{10} = \frac{9}{10}$, '5 tenths plus 4 tenths equals 9 tenths'—which is clear from the meaning of a fraction. But this only became clear when the denominators of both fractions were the same number. We found a suitable denominator (10) by multiplying the two denominators 2×5; but any other method of finding a suitable denominator would work. For example, if called upon to add $\frac{2}{11} + \frac{5}{11}$, we notice that the fractions already have the same denominator (11), so that they can be added immediately by adding the numerators, $\frac{2}{11} + \frac{5}{11} = \frac{2+5}{11} = \frac{7}{11}$. Again, if asked to add $\frac{1}{9} + \frac{1}{18}$, we don't need to choose 9×18 as a suitable denominator for our equivalent fractions (although that

1 mile

$\frac{1}{2}$ mile

$\frac{2}{5}$ mile

$\frac{1}{2} + \frac{2}{5} = \frac{9}{10}$

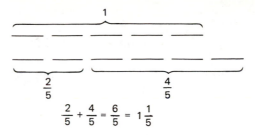

$$\frac{2}{5} + \frac{4}{5} = \frac{6}{5} = 1\frac{1}{5}$$

choice would work). Since 18 is exactly divisible by 9, we can choose 18 itself as a suitable denominator, replace $\frac{1}{9}$ by the equivalent fraction $\frac{2}{18}$, so that

$$\frac{1}{9} + \frac{1}{18} = \frac{2}{18} + \frac{1}{18} = \frac{3}{18}.$$

Notice here that we may now replace $\frac{3}{18}$ by the equivalent fraction $\frac{1}{6}$, so that the neatest answer is

$$\frac{1}{9} + \frac{1}{18} = \frac{1}{6},$$

though the earlier answer $\frac{1}{9} + \frac{1}{18} = \frac{3}{18}$ is, of course, just as true.

There is an important respect in which our original example, involving us in the addition $\frac{1}{2} + \frac{2}{5}$, was special. It may happen that, in adding two proper fractions, the result is an improper fraction, or (in particular) the number 1. For example

$$\frac{2}{5} + \frac{4}{5} = \frac{6}{5}, \text{ and } \frac{6}{5} \text{ is bigger than } 1; \quad \frac{3}{4} + \frac{1}{4} = \frac{4}{4} = 1.$$

How does this affect our rule for adding fractions? If we have to add $17\frac{2}{5} + 29\frac{4}{5}$ we apparently get $46\frac{6}{5}$. This is not, however, a good form in which to give the answer since it is neither in the form of a mixed number (the fractional part is not proper) nor of an improper fraction. To convert our answer to a good form, we simply make the following calculation:

$$46\frac{6}{5} = 46 + \frac{6}{5} = 46 + 1 + \frac{1}{5} = 47 + \frac{1}{5} = 47\frac{1}{5}.$$

so that the answer is $47\frac{1}{5}$.

Thus, to sum up, we give the rule for adding two mixed numbers. But first, a word of advice. The rule, because it is general, and because it covers all possible difficulties that might arise, must contain several different letters. If that makes you feel uncomfortable simply replace all of the M's by some number, say 12, all of the N's by another number, say 5, etc. Remember that A should be less than B, and C should be less than D. Thus you might choose $A = 2$, $B = 3$, $C = 4$ and $D = 5$. Now,

To Add Two Mixed Numbers

Suppose we wish to add

$$M + \frac{A}{B}, \quad N + \frac{C}{D}$$

Example

$$M = 12, A = 2, B = 3, N = 5, C = 4, D = 5$$

so that our numbers are

$$12\frac{2}{3}, \quad 5\frac{4}{5}$$

which is the same as

$$12 + \frac{2}{3}, \quad 5 + \frac{4}{5}$$

(1) add $M + N$;

(1) add $12 + 5 = 17$;

(2) find fractions $\frac{X}{Z}, \frac{Y}{Z}$ equivalent to

$\frac{A}{B}, \frac{C}{D}$ respectively with the *common*

denominator Z ;

(2) $\frac{2}{3} = \frac{10}{15}, \frac{4}{5} = \frac{12}{15}$;

(3) add the fractions $\frac{X}{Z} + \frac{Y}{Z}$ to get

$\frac{X + Y}{Z}$;

(3) $\frac{10}{15} + \frac{12}{15} = \frac{10 + 12}{15} = \frac{22}{15}$;

(4) if $X + Y$ is less than Z, the

final answer is $(M + N) + \dfrac{X + Y}{Z}$;

if $X + Y = Z$, the final answer
is $(M + N + 1)$;

if $X + Y$ is bigger than Z, write

$\dfrac{X + Y}{Z}$ as $1 + \dfrac{W}{Z}$ and the final

answer is $(M + N + 1) + \dfrac{W}{Z}$;

(4) ($10 + 12$ is *not* less than 15)

($10 + 12$ is *not* equal to 15)

$\dfrac{10 + 12}{15} = \dfrac{22}{15} = 1 + \dfrac{7}{15}$;

so $12\frac{2}{3} + 5\frac{4}{5} = 17 + 1 + \dfrac{7}{15} = 18\dfrac{7}{15}$;

(5) note that it may be possible to
reduce the fractional part of the
answer in step (4), and this may
be worth doing.

(5) $\frac{7}{15}$ is already in reduced form.

Notice that if, in step (3) the fraction $\frac{X + Y}{Z}$ is not proper, then it must be less than 2. For X and Y are both less than Z, so $X + Y$ is less than $Z + Z$ or $2 \times Z$. So when we express $\frac{X + Y}{Z}$ as a mixed number the whole number part must be 1.

If you made the suggested substitutions you will already have added $12\frac{2}{3} + 5\frac{4}{5}$ to obtain $18\frac{7}{15}$. You should practice this rule with other numbers, so that it will become routine.† You should then be able to see quite easily how to add together more than two mixed numbers. An example may help to clarify the method.

Example   Add $12\frac{2}{3} + 5\frac{4}{5} + 2\frac{3}{4}$.

First $12 + 5 + 2 = 19$. Next we seek fractions equivalent to $\frac{2}{3}, \frac{4}{5}, \frac{3}{4}$ respectively and all having a common denominator. Now $3 \times 5 \times 4 = 60$, so 60 will serve as a common denominator. We next observe, using Rule 4, that

$$\frac{2}{3} = \frac{2 \times 20}{3 \times 20} = \frac{40}{60}, \frac{4}{5} = \frac{4 \times 12}{5 \times 12} = \frac{48}{60}, \frac{3}{4} = \frac{3 \times 15}{4 \times 15} = \frac{45}{60},$$

thus

$$\frac{2}{3} + \frac{4}{5} + \frac{3}{4} = \frac{40}{60} + \frac{48}{60} + \frac{45}{60} = \frac{133}{60} = 2\frac{13}{60}.$$

Finally, we see that

$$12\frac{2}{3} + 5\frac{4}{5} + 2\frac{3}{4} = 19 + 2\frac{13}{60} = 21\frac{13}{60}.$$

If we are actually given two improper fractions, $\frac{A}{B}, \frac{C}{D}$, to add (as we might be on a test!) then it is perfectly correct to get an improper fraction as answer by just using steps (2) and (3) of the procedure described above. Of course we no longer have A less than B and C less than D, but this doesn't matter. In future we will suppose that we can add mixed numbers or improper fractions.

We should now say a word about subtracting fractions. Here again, we may illustrate the subtraction of fractions by pretended 'real-life' situations involving measures (lengths, weights, etc.) which, for some reason, we happen to have expressed as fractions. Thus we will always suppose that, in subtracting fractions, we are always taking a smaller number from a larger. If our two numbers are given as mixed numbers, $M + \frac{A}{B}$ and $N + \frac{C}{D}$, then it will always be the case that $M \geqslant N$ (M is greater than or equal to N), but a difficulty may occur in tackling the fractional parts, which is illustrated in the

---

†You may wish to do Problem 1 of Exercise 2.5 now—and then return to the text.

problem $7\frac{2}{9} - 4\frac{5}{6}$. Now $7 - 4 = 3$, so we must calculate $\frac{2}{9} - \frac{5}{6}$. Here we find the common denominator 18 (remember that there will be other possible common denominators—you may have thought of 54—we have chosen the smallest). Then $\frac{2}{9} = \frac{4}{18}$, $\frac{5}{6} = \frac{15}{18}$, so we must subtract $\frac{4}{18} - \frac{15}{18}$. But 15 is bigger than 4—what should we do? Now remember that we have met a similar problem in subtracting whole numbers. If we wish to subtract $33 - 17$, we try to subtract 7 from 3; since we cannot, we 'borrow a ten',† that is, take 10 from 30, making it 20 and put this 10 with the units obtaining 13 units. Then we subtract 7 from 13, getting 6, and proceed to subtract 10 from 20. So here we borrow, but this time we borrow a unit. Remember we have already calculated $7 - 4 = 3$, so we have 3 available as a whole number. We regard 3 as $2\frac{18}{18}$ so that

$$7\frac{2}{9} - 4\frac{5}{6} = 3 + \frac{2}{9} - \frac{5}{6} = 2 + \frac{18}{18} + \frac{4}{18} - \frac{15}{18} = 2\frac{7}{18},$$

since $18 + 4 - 15 = 7$. Of course, this difficulty may not arise! The problem $12\frac{1}{2} - 8\frac{1}{3}$ may be dealt with in a straightforward fashion as follows,

$$12\frac{1}{2} - 8\frac{1}{3} = 4 + \frac{1}{2} - \frac{1}{3} = 4 + \frac{3}{6} - \frac{2}{6} = 4\frac{1}{6}.$$

* We close this section with some theoretical remarks about the addition and subtraction of fractions, which the practically-minded reader may wish to simply ignore. (That reader should then also ignore Section 5 of Chapter 3!)

Given any two fractions, $\frac{A}{B}$ and $\frac{C}{D}$, we can always add them without converting them to mixed numbers if we wish, and we may always choose $B \times D$ for a common denominator. (This is, indeed, how a machine would be likely to proceed!) Then

$$\frac{A}{B} = \frac{A \times D}{B \times D}, \qquad \frac{C}{D} = \frac{B \times C}{B \times D},$$

so

$$\frac{A}{B} + \frac{C}{D} = \frac{(A \times D) + (B \times C)}{B \times D}.$$

This is an absolutely general formula for the addition of two fractions. Similarly, if $A/B \geqslant C/D$ (which means, equivalently, that $A \times D \geqslant B \times C$) then we have the general formula for subtracting fractions

$$\frac{A}{B} - \frac{C}{D} = \frac{(A \times D) - (B \times C)}{B \times D}.$$

†Some people think it is more honest to call this process 'stealing' since the quantities are never returned. Others think of it as 'trading in'.

We emphasize, however, that these rules are not likely to be the best procedures for any particular example of adding or subtracting fractions which we want to work out.

Exercise 2.5

1. Perform the following computations

(a) $\dfrac{2}{9} + \dfrac{4}{9} =$

(g) $\dfrac{5}{6} + \dfrac{7}{15} =$

(b) $\dfrac{2}{15} + \dfrac{7}{15} =$

(h) $2\dfrac{3}{4} + 5\dfrac{3}{4} =$

(c) $\dfrac{5}{18} + \dfrac{11}{18} =$

(i) $6\dfrac{7}{8} + 4\dfrac{5}{8} =$

(d) $\dfrac{1}{3} + \dfrac{1}{6} =$

(j) $200\dfrac{1}{15} + 100\dfrac{2}{5} =$

(e) $\dfrac{1}{5} + \dfrac{7}{30} =$

(k) $3\dfrac{2}{3} + 12\dfrac{4}{5} =$

(f) $\dfrac{1}{4} + \dfrac{5}{6} =$

(l) $102\dfrac{7}{15} + 103\dfrac{11}{14} =$

2. Perform the following additions. If you pay attention to the answers (and don't make any errors) you may 'feel' the results are correct without checking them against the ones in the back of the book.

(a) $2\dfrac{2}{9} + 3\dfrac{2}{3} + 4\dfrac{1}{9} =$

(b) $4\dfrac{1}{6} + 5\dfrac{1}{3} + 10\dfrac{1}{2} =$

(c) $15\dfrac{2}{7} + 10\dfrac{1}{3} + 4\dfrac{8}{21} =$

(d) $20\dfrac{20}{33} + 10\dfrac{2}{3} + 8\dfrac{8}{11} =$

3. Perform the following additions and express your final answer in *decimal* form. (This is not possible, in general, but these problems have been constructed so that it can be done.)

(a) $3\dfrac{5}{7} + 10\dfrac{4}{7} + 1\dfrac{3}{14} =$

(d) $10\dfrac{2}{3} + 8\dfrac{3}{4} + 2\dfrac{5}{6} =$

(b) $3\dfrac{1}{5} + 2\dfrac{1}{10} + 4\dfrac{1}{4} =$

(e) $6\dfrac{1}{20} + 2\dfrac{1}{5} + 7\dfrac{3}{4} =$

(c) $5\dfrac{2}{5} + 2\dfrac{1}{2} + 4\dfrac{1}{25} =$

(f) $5\dfrac{3}{5} + 6\dfrac{13}{20} + 2\dfrac{16}{25} =$

4. Notice that Problem 3 (b) could have been computed by observing that

$$\dfrac{1}{5} = \dfrac{2}{10} = .2,$$

$$\frac{1}{10} = .1,$$

$$\frac{1}{4} = \frac{25}{100} = .25,$$

and then adding the decimal numbers as follows

$$
\begin{array}{r}
3.2 \\
2.1 \\
+\ 4.25 \\
\hline
9.55
\end{array}
$$

Do parts (c), (e) and (f) of Problem 3 by this method, if you didn't already think of doing that. If you did, congratulations!

5.   A Stable Fable

Once upon a time there was a dying man who owned 17 horses. In his will he decreed that his oldest son should receive $\frac{1}{2}$ of the horses, the middle son $\frac{1}{3}$ of the horses, and the youngest son $\frac{1}{9}$ of the horses. Now, his sons were some of the pioneers of 'math avoidance' and consequently they had no idea how to divide up the horses. In desperation they called in the local mathematics professor who arrived riding his own fine horse. The professor dismounted and confidently tied his horse next to the other 17 horses in the stable. He proceeded, with lightning speed, to solve the problem that had so vexed the three sons. 'It's simple,' he said. 'Actually, in fact, it's quite obvious.' And, as mathematics professors sometimes did in those days, he pronounced *without explanation* that

$$\frac{1}{2} \text{ of 18 is 9,}$$

$$\frac{1}{3} \text{ of 18 is 6,}$$

$\frac{1}{9}$ of 18 is 2,

so the oldest son should receive 9 horses, the middle son 6 horses, and the youngest son 2 horses.'

Having solved the problem the professor got on his own horse and rode away. Of course the sons were pleased because they *knew* that $9 + 6 + 2 = 17$.

(a)   How many horses should the old man have had if the fractions $\frac{1}{2}$, $\frac{1}{4}$, $\frac{1}{6}$ would have worked in this story? (Hint: Find $\frac{1}{2} + \frac{1}{3} + \frac{1}{9}$.)

(b)   Find another set of fractions and a corresponding number of horses that would have worked in this story. (There are many possibilities.)

6.   Perform the following subtractions.

(a) $5\frac{3}{4} - 3\frac{1}{4}$ (d) $11\frac{1}{4} - 5\frac{3}{8}$

(b) $11\frac{3}{4} - 5\frac{2}{3}$ (e) $20\frac{1}{3} - 10\frac{3}{7}$

(c) $21\frac{7}{8} - 13\frac{4}{5}$ (f) $142\frac{1}{10} - 72\frac{3}{10}$

7.   A $27\frac{3}{8}$ inch length of two-by-four is needed; it is to be cut from a $36\frac{1}{4}$ inch piece. How long would the remaining piece be if the saw in question makes a cut $\frac{1}{8}$ inch wide?

2.6 PROBABILITY THEORY

We will discuss here situations in which an experiment is conducted which has *finitely* many possible outcomes. An example of this is the toss of a coin, where there are 2 possible outcomes (heads or tails); another is the throw of a die, where there are 6 possible outcomes (1, 2, 3, 4, 5, 6); another is the selection of a student in a class, where the number of outcomes is the number of students in that class. A *counterexample* (that is, an example of an experiment with *infinitely* many possible outcomes) is to say to a person, 'Think of a number'.

We assume that each of the outcomes is equally likely. So let us suppose that, for our experiment, there are D possible outcomes. Suppose further that we are interested in some quality, some feature, that each of the outcomes may, or may not, possess. If N of the outcomes have that feature, then we say that the *probability* of our experiment producing that feature is the fraction† $\frac{N}{D}$. We will give two examples.

---

†It would be more correct, but fussy, to say that the probability is the rational number represented by the fraction $\frac{N}{D}$.

Let us throw a die and suppose we win if the number showing up on the die is odd. Then, of the 6 possible outcomes, we win in 3 cases, namely, if the number 1, 3 or 5 shows up. Thus the probability of winning is the fraction $\frac{3}{6}$ or $\frac{1}{2}$.

Next, suppose in a class of 37 children, there are 20 boys and 17 girls. A child is to be chosen to represent the class at a ceremony. Then the probability that a girl will be chosen (if the selection is a random one) is $\frac{17}{37}$.

The feature of an experiment that is of interest to us may, of course, be one particular outcome. In that case, if there are D possible (and equally likely) outcomes, the probability of that particular outcome is $\frac{1}{D}$. For example, the probability of a coin coming down heads is $\frac{1}{2}$.

It is important to remember that the possible outcomes are to be equally likely. In our second example above one could think of the 'outcomes' of the experiment as 'A boy is selected', 'A girl is selected', and one might be led to the false conclusion that the probability that a girl is selected is $\frac{1}{2}$; this would be false because, in view of the composition of the class, the two 'outcomes' are not equally likely.

The point we are making may seem obvious but it led a famous thinker, D'Alembert, to make a serious mistake which would have been ruinous to a gambler. He considered the possible outcomes of the experiment of tossing *two* coins and listed them as:

A: both coins show heads
B: both coins show tails
C: one coin shows heads, the other shows tails

He inferred that the probability of outcome C is $\frac{1}{3}$, but this is wrong, since the outcomes A, B, C are not equally likely. If we call the coins X and Y, then we may list 4 possible *and equally likely* outcomes:

X shows a head; Y shows a head
X shows a head; Y shows a tail
X shows a tail; Y shows a head
X shows a tail; Y shows a tail

The feature which D'Alembert favored (one coin showing a head, the other a tail) occurs in 2 of these 4 outcomes, so its probability is $\frac{2}{4}$ or $\frac{1}{2}$, not $\frac{1}{3}$.

Why is the notion of probability important? It is because it tells you what to expect in repeated trials of an experiment. If we toss a coin 100 times, the number of times it shows heads will be approximately $\frac{1}{2}$ of 100, or 50, since the probability of heads in a single trial is $\frac{1}{2}$. If we throw a die 60 times, the number of times it shows 5 will be approximately $\frac{1}{6}$ of 60, or 10, since the probability of 5 showing up in a single throw is $\frac{1}{6}$. Thus probability theory tells us what to expect, and thus enables us to behave rationally.

Probabilities then are measured by fractions and we would expect the arithmetic of fractions to help us in the development of probability theory. Thus it is natural to ask: what meaning do the addition and multiplication of fractions have in probability theory? Before answering this question, we remark that the fractions that come into probability theory are all proper fractions. More precisely all such fractions lie between 0 and 1. This is because we look at *all* possible outcomes (whose number is the denominator of our fraction) and then select only those we favor (whose number is the numerator of our fraction). The extreme cases of probability 0 and probability 1 do occur. So long as we confine our attention to a *finite* number of possible outcomes, then probability 0 means that a favorable outcome is impossible (the probability of 8 showing up on our die); and probability 1 means that a favorable outcome is certain (the probability of a number less than 7 showing up on our die).

Let us first discuss the multiplication of fractions. We have already seen one example of this principle. The probability of heads coming up in a single toss of a coin is $\frac{1}{2}$. If we ask what is the probability of heads coming up on two coins then we may deduce from the following illustration, which shows all of the possibilities, that the answer is $\frac{1}{4}$.

Let us now give another example. If we throw two dice, what is the probability that the first die shows 1 or 2, and that the second die shows 1 or 2 or 3? The probability that a single die shows 1 or 2 is $\frac{2}{6}$ or $\frac{1}{3}$; the probability that a single die shows 1 or 2 or 3 is $\frac{3}{6}$ or $\frac{1}{2}$. And we see from the following table that the number of cases where the first die is a 1 or 2 and the second is a 1 or 2 or 3 is 6 (or 2×3), and the total number of possible outcomes is 36 (or 6×6).

Second die \ First die	1	2	3	4	5	6
1	1 1	2 1	3 1	4 1	5 1	6 1
2	1 2	2 2	3 2	4 2	5 2	6 2
3	1 3	2 3	3 3	4 3	5 3	6 3
4	1 4	2 4	3 4	4 4	5 4	6 4
5	1 5	2 5	3 5	4 5	5 5	6 5
6	1 6	2 6	3 6	4 6	5 6	6 6

Thus the required probability is $\frac{6}{36}$ or $\frac{1}{6}$. But we would not wish to carry out such a tedious listing for every problem of this type so we look for a reasonable way to get each of these answers from the probabilities of the separate events. We observe in the first example that the answer $\frac{1}{4}$ is the same as $\frac{1}{2} \times \frac{1}{2}$; and, in the second example the answer $\frac{1}{6}$ is the same as $\frac{1}{3} \times \frac{1}{2}$. It seems very likely that these two examples represent a more general concept. Indeed this is the case, and it can be supported by the following general argument.

Suppose we consider two quite distinct experiments. The first has B possible outcomes, of which A are favorable; the second has D possible outcomes, of which C are favorable. Then the probability of a favorable outcome in the first experiment is $\frac{A}{B}$; and the probability of a favorable outcome in the second experiment is $\frac{C}{D}$.

Now, let us regard the two experiments together as a single experiment. Then we may regard, as an outcome of this single composite experiment, a pair of outcomes, one of the first experiment, the other of the second experiment. How many such pairs of outcomes are there? Plainly there are $B \times D$ pairs (see the table for the two dice, as an example). How many of those $B \times D$ outcomes are favorable? For a composite outcome to be favorable, each of the two individual outcomes must be favorable. Thus the number of favorable composite outcomes is $A \times C$. We conclude that the probability of a favorable outcome in the 'joint' experiment is

$$\frac{A \times C}{B \times D}, \text{ or } \frac{A}{B} \times \frac{C}{D}.$$

Thus we have found a natural and important interpretation of the product of two fractions.

To illustrate the utility of this idea let us return to the example considered earlier of the class of 37 children, of whom 20 are boys and 17 are girls. Suppose we are told also that, of the 37 children, 23 are dark-haired and 14 are light-haired. How many dark-haired girls should we expect to find in the class? We could argue as follows. The probability that a child is female is $\frac{17}{37}$ and the probability that a child is dark-haired is $\frac{23}{37}$. Thus, assuming that the two properties of a child, that of being female and that of being dark-haired, are independent, we deduce that the probability that a child is *both* female and dark-haired is $\frac{17}{37} \times \frac{23}{37}$. We therefore expect $\frac{17}{37} \times \frac{23}{37} \times 37$ (or approximately 11) female, dark-haired children in the class. Of course, this is only an estimate, but it is the best estimate we can make on the information we have. It is fair to describe this number 11 as our *expectation,* on the evidence we have, of the number of female, dark-haired children in the class. This estimate can be used in the following very significant way. Suppose we now do a count and find that there are, in fact, 15 female, dark-haired children in the class. Since 15 is decidedly bigger than 11, we infer that there must be a (*positive*) *correlation* between the properties of being female and of being dark-haired; that is, those two properties seem not to be independent, but rather to tend to 'go together'. Similarly, if the actual number of female, dark-haired children had been only 6, we would have inferred a *negative* correlation between those two properties.† This then gives us a key technique in a scientific method of inquiry—we look for statistical correlations between properties and, if we find them, we then look for possible causal connections. You will all be familiar with this method in connection with the relation of cigarette-smoking and lung cancer.

We now discuss how the *addition* of fractions enters into probability theory. First, we give an example. Let us throw 2 dice and suppose you win if both dice come up odd or both dice come up even. The probability that both dice come up odd is $\frac{1}{2} \times \frac{1}{2} = \frac{1}{4}$. Similarly, the probability that both dice come up even is $\frac{1}{2} \times \frac{1}{2} = \frac{1}{4}$. Now the probability that either both dice come up odd or both dice come up even is seen, from our table, to be $\frac{9+9}{36}$ or $\frac{1}{2}$ (which is also $\frac{9}{36} + \frac{9}{36}$ or $\frac{1}{4} + \frac{1}{4}$).

Now consider a slightly different type of example. Suppose you throw two dice and win if the sum of the two dice is either less than or equal to 5 or greater than or equal to 8. What is the probability of your winning? From our table concerning two dice we see that

---

†We emphasize that such inferences are provisional; further evidence might cause us to modify them, or even abandon them.

the probability the sum of two
dice is less than or equal to 5 $= \dfrac{10}{36}$,

the probability the sum of two
dice is greater than or equal to 8 $= \dfrac{15}{36}$,

But we can also see from the table that the answer to our question is $\frac{25}{36}$, which is $\frac{10}{36} + \frac{15}{36}$. As we have seen so many times before, these specific examples suggest a generalization. We now formalize that important idea.

Suppose we conduct an experiment and we are interested in two features, F and G, which the possible outcomes might present. Suppose further that *no outcome could have both of these features,* and we ask for the probability that an outcome has *either* feature F *or* feature G. Now if there are D possible outcomes and if A of them have feature F, while B of them have feature G, then there are exactly $A + B$ outcomes having either F or G. Thus

the probability of F is $\dfrac{A}{D}$,

the probability of G is $\dfrac{B}{D}$,

the probability of F or G is $\dfrac{A + B}{D} = \dfrac{A}{D} + \dfrac{B}{D}$.

Thus we *add* the fractions giving the probabilities of F and G to obtain the probability of F or G.

You should observe here that, in general, it is *not* wise to reduce the fractions that represent the probabilities of F and G, respectively. If you do reduce those fractions it may make the subsequent addition more difficult. In the last example, for instance, if we had reduced the fractions we would have had to compute $\frac{5}{18} + \frac{5}{12}$ instead of $\frac{10}{36} + \frac{15}{36}$. The former is clearly more difficult. The lesson to be learned from this example is that unless you have a reason for reducing a number representing a probability, it is wisest to leave it in its original form.

As a last example of this concept, suppose that in a zoo of 100 animals, there are 8 elephants and 3 lions. What is the probability that the first animal you see will be an elephant or a lion? In the absence of any further information (about the relation of the usual locations of the elephants and lions to the entrance to the zoo, about the shyness of animals, etc.) the required probability is

$$\frac{8}{100} + \frac{3}{100} = \frac{11}{100}.$$

Remark

In case you feel you may have some difficulty in remembering, in probability problems, whether the fractions involved have to be multiplied or added, it may be helpful to point out that when we are dealing with a *compound* event (the probability of both event *A* *and* event *B*) we multiply probabilities; and when we are dealing with *disjoint* events (the probability of either event *A* *or* event *B*) we add probabilities. A good mnemonic is "and = multiply, or = add". However, we should give you two warnings, of a very different nature the one from the other. First, it can be *very dangerous* to try to remember something you don't understand. Second, 'or' only corresponds to 'add' if the events in question are *mutually exclusive*. For example, if a baby was born during the month of February, 1980, the probability that it was born on a Friday is $\frac{5}{29}$ (why?); but the probability that it was born on February 1 or on a Friday is also $\frac{5}{29}$ (why?).

Exercise 2.6

1. If you roll a single die twice, what is the probability that the first time the die will show a 1 or a 2 and the second time it will show a 1 or 2 or 3? (Compare your answer with the example in the text where two dice were thrown.)

2. What is the probability that when you roll a single die twice you will roll
    (a)   a 1 first and a 2 second?
    (b)   a 2 first and a 3 second?
    (c)   a 3 first and a 4 second?
    (d)   a 4 first and a 5 second?
    (e)   a 5 first and a 6 second?
    (f)   two consecutive numbers in increasing order?

3. If a student handed in a solution to Problem 2(f) as follows

$$\frac{5}{6} \times \frac{1}{6} = \frac{5}{36} \, ,$$

should he or she receive credit for the answer? (Is there any way to argue that the probability of the first die landing successfully is $\frac{5}{6}$ and that then the probability that the second die lands successfully is $\frac{1}{6}$?)

4. (a) If you throw two dice what is the probability that both dice will show the same number?

    (b) If you throw three dice what is the probability that all three will show the same number?

5. If you roll a single die two times what is the probability the first number will be less than the second? (Hint: Look at the listing for the pair of dice if necessary.)

6. If a student turned in a solution to Problem 5 as follows

$$\frac{1}{6} \times \frac{5}{6} + \frac{1}{6} \times \frac{4}{6} + \frac{1}{6} \times \frac{3}{6} + \frac{1}{6} \times \frac{2}{6} + \frac{1}{6} \times \frac{1}{6} =$$

$$\frac{5}{36} + \frac{4}{36} + \frac{3}{36} + \frac{2}{36} + \frac{1}{36} = \frac{15}{36} = \frac{5}{12} \, ,$$

should he or she receive credit for it? (Hint: Think of the five different successful possibilities for the first die.)

7. If you roll a single die two times what is the probability the first number will not be less than the second?

8. If a student turned in a solution of problem 7 as follows

$$1 - \frac{5}{12} = \frac{7}{12} \, ,$$

should he or she receive credit for it? (Hint: Recall that it is certain that either an event happens or it doesn't. You already know from Problem 5 the probability that the first number *is* less than the second.)

2.7 ODDS, PROBABILITY, AND EXPECTATION

In this section we explain the relationship between the terms 'probability' and 'odds' and introduce an important scientific result relating to experiments and how their outcomes change the odds in favor of a given hypothesis.

Gamblers are more prone to express the chance of some event (such as a horse winning a race) in terms of odds rather than of probabilities. The two concepts are, however, very closely related, and each may be obtained from the other. The *probability* of a favorable outcome is, as we have said, the ratio

$$\frac{number\ of\ favorable\ outcomes}{total\ number\ of\ outcomes} \, , \quad or \quad \frac{N}{D} \, ,$$

where N is the number of favorable outcomes and D is the total number of outcomes. Then $D - N$ is the number of *unfavorable* outcomes and we call the ratio $\frac{N}{D-N}$ of favorable to unfavorable outcomes the *odds* on a favorable outcome. (Actually if the ratio is less than 1 we often use the ratio $\frac{D-N}{N}$ which we call, quite naturally, the odds *against* a favorable outcome). Thus the odds *against* throwing a 4 with a die are 5 to 1. The odds *on* a head turning up at least once when we toss two coins are 3 to 1. Notice that odds are usually (that is, traditionally) announced as ratios (3 to 1, 5 to 2, etc.) rather than as fractions ($\frac{3}{1}$, $\frac{5}{2}$, etc.).

There is quite clearly a connection between the odds and the probability of an outcome for a given experiment. From the information in the first example above we could infer that the die has (1 + 5 =) 6 faces and 5 of them are not 4's. Thus the probability of *not* throwing a four with a single die is $\frac{5}{1+5}$ (or $\frac{5}{6}$). In the second example we could conclude that if there were (1 + 3 =) 4 cases then 3 of them would be favorable so that the probability of a head being up at least once when we toss two coins is $\frac{3}{1+3}$ (or $\frac{3}{4}$). If we had a similar experiment in which the odds for a favorable event are 5 to 2, then we might conclude that there are (2 + 5 =) 7 possible outcomes of which 5 have the favorable feature. Thus the probability for the favorable occurrence is $\frac{5}{2+5}$ (or $\frac{5}{7}$).

Of course it goes the other way too. That is, if you know the probability that some event will occur you can determine the odds. Thus, for example, if you throw a single die, the probability you will throw a number bigger than or equal to 3 is $\frac{4}{6}$. It follows that out of 6 cases 4 are favorable and therefore 2 are not favorable, so the odds in favor of this event are 4 to 2. Notice that if you had reduced the fraction representing the probability to $\frac{2}{3}$ then you would have been thinking of 3 cases of which 2 are favorable and 1 is not favorable. You would have then obtained the odds of 2 to 1 (which, of course, is equivalent to the previous answer).

We now wish to establish the connection between the probability P and the odds Q. You may wish to skip the argument and just note the final results stated as Rule 6.

* As you will see these results are obtained by using ratios that have fractions in the numerator and fractions in the denominator. Thus we begin to suspect that it may be necessary to learn how to divide fractions by fractions. We will delay the details, however, until the next chapter, being content here to simply use the concepts as motivation for that procedure. The exercises (except for the last) will be constructed so that you will not be asked to divide fractions by fractions.

Let us proceed by recalling that

$$P = \frac{N}{D} \qquad \text{and} \qquad Q = \frac{N}{D-N},$$

where D is the total number of outcomes and N is the number of favorable outcomes. Then by Rule 4 we can multiply the top and bottom of $\frac{N}{D-N}$ by $\frac{1}{D}$ to obtain

$$Q = \frac{N \times \frac{1}{D}}{(D-N) \times \frac{1}{D}} = \frac{\frac{N}{D}}{\left(D \times \frac{1}{D}\right) - \left(N \times \frac{1}{D}\right)} = \frac{\frac{N}{D}}{1 - \frac{N}{D}} ;$$

and since $\frac{N}{D} = P$ we have

$$Q = \frac{P}{1-P} .$$

This is a *mathematical formula* relating P and Q, enabling you to obtain the value of the odds Q from that of the probability P by means of some standard manipulation of fractions. Notice that you don't need to know N and D to obtain Q from P. However if P is expressed as a fraction $\frac{A}{B}$ (it is not necessary that the fraction actually be $\frac{N}{D}$; we might well have reduced $\frac{N}{D}$ to $\frac{A}{B}$, then the *practical rule* for obtaining a fraction representing Q is to retain the numerator A and to replace the denominator by $B - A$; thus

$$\text{if } P = \frac{A}{B}, \qquad Q = \frac{A}{B-A} .$$

We may proceed similarly to find a mathematical formula and a practical rule for obtaining P from Q. To find the mathematical formula, we 'solve' the equation $Q = \frac{P}{1-P}$ for P. Thus we first multiply by $1 - P$ to obtain

$$Q \times (1 - P) = P.$$

We infer that

$$Q - (Q \times P) = P,$$

so that

$$(Q \times P) + P = Q,$$

or

$$(Q + 1) \times P = Q,$$

and, finally,

$$P = \frac{Q}{Q+1} .$$

The practical rule is simple and more useful in concrete cases. If Q is expressed as a fraction $\frac{X}{Y}$, then we obtain a fraction representing P by retaining the numerator X and replacing the denominator by $X + Y$; thus

If $Q = \dfrac{X}{Y}$, $P = \dfrac{X}{X+Y}$.

It is often natural, especially if our 'experiment' is a game, to think in terms of winning and losing rather than favorable and unfavorable outcomes. Suppose then that of the D possible outcomes of our game, N result in our winning and $D - N$ result in our losing. Then if P is the probability of winning, $\overline{P}$ the probability of losing, Q the odds on winning, and $\overline{Q}$ the odds on losing, we have

$$P = \frac{N}{D}, \qquad \overline{P} = \frac{D-N}{D}, \qquad Q = \frac{N}{D-N}, \qquad \overline{Q} = \frac{D-N}{N}.$$

Notice the important relations

$$P + \overline{P} = 1, \qquad Q \times \overline{Q} = 1.$$

Summing up, we have the following rule.

Rule 6   Let the probability of an event be P and the odds on the event Q.

(a)   If P (probability) is the fraction $\dfrac{A}{B}$, then Q (odds) is the fraction $\dfrac{A}{B-A}$.

(b)   If Q (odds) is the fraction $\dfrac{X}{Y}$, then P (probability) is the fraction $\dfrac{X}{X+Y}$.

As examples, first suppose that the probability of an event is $\frac{2}{3}$. Then the odds are $\frac{2}{3-2}$ or $\frac{2}{1}$ ('two to one')† Notice that we may describe the probability by the fraction $\frac{6}{9}$, equivalent to $\frac{2}{3}$. But then our rule for the odds yields the fraction $\frac{6}{9-6}$ or $\frac{6}{3}$, which is equivalent to $\frac{2}{1}$, as it should be!

Going in the other direction suppose that the odds are $\frac{2}{5}$ ('two to five'). Then the probability is $\frac{2}{2+5}$ or $\frac{2}{7}$.

We may well meet a generalized version of the situation described at the beginning of the section, in which the set of outcomes is partitioned into more than two subsets. We illustrate this in the following example.

Example I   Suppose you play a game in which you roll a single die. This particular game has the following possible results:

Result 1. If you roll a 2 you win $2.00 (and recover your stake).

---

†It is customary to express probabilities as fractions ('three fifths', 'two sevenths', etc.) and the odds as ratios ('two to three', 'two to five', etc.). This distinction is, of course purely conventional; it is perhaps due to the fact that probabilities are much used by scientists and engineers, while odds are much used by gamblers.

Result 2. If you roll a 4 or 6 you win $6.00 (and recover your stake).

Result 3. If you roll a 1, 3 or 5 you lose your stake.

The question is: What amount S (your $Stake$) should you pay each time you play if the game is going to be 'fair'?†

Solution   We proceed systematically (so that you may be able to reproduce the reasoning for other similar games). First the results could be labeled,

R_1 = a win of $2.00 = +2,
R_2 = a win of $6.00 = +6,
R_3 = a loss of S(dollars) = $-S$.

Now the number identified with R_k tells us how many dollars you would be ahead, or would have lost, if you achieved the kth result.

Next we identify the corresponding probabilities connected with each result. Thus we record that

the probability P_1 of winning $2.00 ($R_1$ = +2) by rolling a 2 is $\frac{1}{6}$ ($P_1 = \frac{1}{6}$),

the probability P_2 of winning $6.00 ($R_2$ = +6) by rolling a 2 or 4 is $\frac{2}{6}$ ($P_2 = \frac{2}{6}$),

the probability P_3 of losing S dollars ($R_3 = -S$) by rolling a 1, 3 or 5 is $\frac{3}{6}$ ($P_3 = \frac{3}{6}$).

This means that, in the long run,

you win $2.00 one sixth of the time,
you win $6.00 two sixths of the time,
you lose S dollars three sixths of the time.

From these statements we can see that your expected gains (or winnings) from Result 1 and Result 2, in the long run, say for n plays, can be expressed as

$$\$2.00 \times \frac{1}{6} \times n + \$6.00 \times \frac{2}{6} \times n.$$

Likewise, your expected losses, in the long run (in fact, for the same number of plays, n) can be expressed by

$$S(\text{dollars}) \times \frac{3}{6} \times n.$$

This last amount can also be expressed by saying that your expected gain from Result 3, in the long run, is

---

†Intuitively you should see that if you play a game a large number of times and if you don't expect to be very much ahead, nor very much 'in the hole,' then the game is a *fair* game. We will give a precise definition later.

$$-S \text{(dollars)} \times \frac{3}{6} \times n.$$

(A negative gain is the same as a loss!)

Now it is possible to say that a game is *fair* in two equivalent ways.

 (a) A game is fair if, in the long run, the net result of summing all of your gains (both positive and negative) is zero.

 (b) A game is fair if, in the long run, the sum of your gains is equal to the sum of your losses.

If we use statement (a) we obtain the equation

$$\underbrace{2 \times \frac{1}{6} \times n + 6 \times \frac{2}{6} \times n + (-S) \times \frac{3}{6} \times n}_{} = 0$$

   sum of the expected gains from
   Results 1, 2 and 3 (remember,
   Result 3 is a negative gain, or loss)

If we use statement (b) we obtain the equation

$$\underbrace{2 \times \frac{1}{6} \times n + 6 \times \frac{2}{6} \times n}_{} = \underbrace{S \times \frac{3}{6} \times n.}_{}$$

  sum of expected gains   sum of expected
  from Results 1 and 2   losses from Result 3

Of course these two equations are equivalent to each other. The second has the slight advantage that we do not need to subtract, or cope with the concept of negative numbers (which will be discussed in Chapter 5).

Dealing with the second equation, first notice that since n appears in every term on both sides of the equals sign, we can divide n into every term on both sides of the equals sign and obtain

$$\left(2 \times \frac{1}{6}\right) + \left(6 \times \frac{2}{6}\right) = S \times \frac{3}{6}. \; ^\dagger$$

Then, multiplying both sides by 6,

$$2 + (6 \times 2) = S \times 3,$$

or

$$14 = S \times 3,$$

---

†In cases where the solution to a problem turns out *not* to depend on one of the variables used we say that the solution is 'independent' of that variable. Thus we have just shown that the solution to our equation is independent of n, the number of times you play the game.

so that

$$\frac{14}{3} = S.$$

This last equation tells us that if you paid $\frac{14}{3}$, or $4\frac{2}{3}$ dollars each time you play this game, and if you play for a large number of times, you would be likely to 'break even' (that is, you won't win, or lose, ver' much).

In practice the gambling house would probably charge $5.00 for each time you play the game in Example I. Consequently the house would make, on the average, $\frac{1}{3}$ of a dollar every time this game is played.

The arguments used in analyzing this example lead us naturally into a very important concept. If we revert to the situation of this example, but argue from the *first* equation, we see that we concluded that the game was fair provided that the quantity

$$E = \left(2 \times \frac{1}{6}\right) + \left(6 \times \frac{2}{6}\right) - \left(S \times \frac{3}{6}\right)$$

is zero. This suggests that, given any game, we define the *expectation* of the game, E, as above, so that we may say that *a game is fair provided that the expectation is zero.* From the arithmetical point of view, there is the minor difficulty that we may have to handle negative numbers, but we feel reasonably sure that those of you who have read this far in this section are comfortable with the amount of the arithmetic of negative numbers involved—if you are not, you could reread this material after studying Section 1 of Chapter 5 (or you could simply insert a study of that section in your reading at this point).

Let us give another example.† Suppose we toss a coin 10 times, and record the sequence of heads and tails; thus a possibility is HTTHHHTHTT. The following are possible results.

Result 1. There are 4 or more successive H's in the sequence; then we lose $1.

Result 2. Result 1 does not occur but there are more heads than tails; then we win $2.

Result 3. Neither Result 1 nor Result 2 occurs; then we lose 10¢.

What is our expectation?

To answer this question, we first note that there are 2^{10} = 1024 possible outcomes (or sequences of H's and T's). *Very* careful counting shows us that Result 1 occurs in 251 cases (this figure surprised us—we thought it would be smaller!). There are 386 cases of more heads than tails, but in 208 of those cases Result 1 occurs. Thus Result 2 occurs in 178 cases. Finally, there are 595 cases of Result 3 (this is just 1024 − (251 + 178)). Thus, in the obvious notation (the same as Example 1, essentially)

$$P_1 = \frac{251}{1024}, \; P_2 = \frac{178}{1024}, \; P_3 = \frac{595}{1024},$$

$$R_1 = -1, \; R_2 = +2, \; R_3 = -0.1.$$

The expectation from this game is therefore given by

$$E = (-1) \times \frac{251}{1024} + (2) \times \frac{178}{1024} + (-0.1) \times \frac{595}{1024}$$

$$= -\frac{251}{1024} + \frac{356}{1024} - \frac{59.5}{1024}$$

$$= \frac{1}{1024} (-251 + 356 - 59.5)$$

$$= \frac{45.5}{1024}$$

$$= +0.04 \text{ (approximately)}$$

This means that the game is nearly but not quite fair; it is favorable to the player. The figure of +0.04 for the expectation means that, on the average, we win 4¢ each time we play. Indeed, the calculation of E can be interpreted as telling us that we expect to win $45.50 if we play the game 1024 times. In general, if we know the expectation E of a particular game, then if we play the game N times, we expect to win the amount $N \times E$ (that is, of course, a loss if E is negative).

---

†This example would be very difficult (and tedious) if we calculated by hand, using simple 'enumeration of cases'. The reader may be content to accept our calculation of P_1, P_2, P_3.

There is a particularly important relationship between the concept of a fair game ($E = 0$) and that of odds, in a game with only two outcomes. Thus let us consider a game in which there are two possible results ('win' and 'lose'). The probability of winning is P and, if we win, we win the amount W. The probability of losing is then $1 - P$ (since we are *certain* to win or lose) and suppose we lose the amount L if we lose. Then our expectation is

$$E = (P \times W) - ((1 - P) \times L).$$

The game is thus fair ($E = 0$) precisely when $P \times W = (1 - P) \times L$, or

$$\frac{P}{1 - P} = \frac{L}{W} .$$

But $\frac{P}{1 - P}$ is just the odds on winning (we called this Q earlier). Thus we have established the principle for a 'win-or-lose' game:

such a game is fair precisely when the odds on winning are equal to the ratio of the amount you lose (if you lose) to the amount you win (if you win).

There are two possibilities if the game is not fair; we may have $Q > \frac{L}{W}$, when the game is favorable to you, or $Q < \frac{L}{W}$ when the game is unfavorable to you. When a gambling house 'gives you odds', this means that the house specifies the ratio $\frac{L}{W}$. Thus if the gambling house (or the bookmaker at the races) gives you odds of† '5 to 2' you must stake $5 to win $2. Thus $L = 5$, $W = 2$; $\frac{L}{W} = \frac{5}{2}$. Thus you see that *a bet is favorable to you precisely when the true odds exceed the odds given by the gambling house or bookmaker.* Be very careful to distinguish between the *true* odds and the odds prescribed by the gambling house or given to you by the bookmaker!

The ideas used in Example 1 generalize in the following way. Suppose that an experiment (it could be anything from a measurement to a gambling game) can lead to k possible numerical results (in the case of a game, you win if the result is positive and you lose if the result is negative). Let those k results be denoted $R_1, R_2, R_3, \cdots, R_k$. Then let the corresponding probabilities of those results occurring be denoted $P_1, P_2, P_3, \cdots, P_k$. We know, since *one* of these results is bound to occur, that

(7.1) $$P_1 + P_2 + P_3 + \cdots + P_k = 1$$

and that the expectation E is given by

$$E = (R_1 \times P_1) + (R_2 \times P_2) + (R_3 \times P_3) + \cdots + (R_k \times P_k).$$

---

†The bookmaker will say '5 to 2 on', and use the phrase '5 to 2 against' for what we call '2 to 5'.

You may verify that equation (7.1) holds, and that $E = 0$ in Example 1, where $R_1 = 2$, $R_2 = 6$, $P_1 = \frac{1}{6}$, $P_2 = \frac{2}{6}$, $P_3 = \frac{3}{6}$ were given, and R_3 was found to be $\frac{-14}{3}$.

We now take another example drawn from the world of gambling.

Example II The game of roulette involves the numbers 0 to 36 (a total of 37 numbers) of which the number 0 is uncolored, the odd numbers between 1 and 36 are colored black (say) and the even numbers between 1 and 36 are colored red. You can choose to bet on red or black and you are then 'given' odds of one to one (called 'evens'!). This means that if you pay $1.00 to play, and then choose red or black, and if your selected color comes up you win $1.00 and recover your stake; otherwise you forfeit your stake. (Thus $L = 1$, $W = 1$ in our earlier notation.) Is this a fair game?

Solution Since there are 18 red and 18 black numbers it makes no difference whether you bet on red or black. Suppose, for the sake of discussion, you bet on red. Then the possible results of your bet are

> *Result 1.* Red comes up and you win $1.00 (and recover your stake).
> *Result 2.* Black or 0 comes up and you lose your stake of $1.00.

We know that the probabilities associated with these results are

$$\frac{18}{37} \text{ that red is the outcome}$$

and

$$\frac{19}{37} \text{ that black or 0 is the outcome.}$$

Of course, it is now obvious that you lose money in the long run, because $\frac{18}{37}$ is less than $\frac{19}{37}$ and the second fraction is the one connected with your losses!

In this example $R_1 = 1$, $R_2 = -1$, $P_1 = \frac{18}{37}$, $P_2 = \frac{19}{37}$.

You should observe that in Example II equation (7.1) is satisfied if we use the probabilities we computed from the description of the game. However the equation $E = 0$ is *not* satisfied, using these same probabilities. Since $(R_1 \times P_1) + (R_2 \times P_2) = (1 \times \frac{18}{37}) + ((-1) \times \frac{19}{37}) = \frac{-1}{37}$, your expectation is negative! And, in fact, the $\frac{-1}{37}$ means that 'on the average' you lose $\frac{1}{37}$ of a dollar (about 2.7¢) every time you stake one dollar on this game.

The point, in Example II, is that you are not getting fair odds. The fair, or true, odds are 18 to 19; the *official* odds, by which we mean the ratio of the amount staked to the amount won (*if* you win) are 1 to 1. Since the official odds are greater than the true odds the game is unfair to you and you would be well advised not to play it!

In a horse race, or a betting situation of a similar kind, it is not possible to determine the true odds. But there is one condition to be satisfied if the given odds are to have a chance of being fair—the associated 'given' probabilities must add up to 1. This will generally be so in a 2-horse race, but only in a 2-horse race. If the odds are announced as, say

4 to 1 on horse *A*,
4 to 1 against horse *B*,

in a 2-horse race, this is intrinsically fair since the associated probabilities are $\frac{4}{5}$ and $\frac{1}{5}$, and the odds imply that, if this race were repeated a large number of times, we would expect horse *A* to win '4 times out of 5' or $\frac{4}{5}$ of the times. In races involving more than 2 horses you will almost always find that if you work out the probabilities associated with the given odds, then the probabilities add up to more than 1. This is unfair to you as the gambler—as we would say, *the odds are against you.* A detailed discussion of this fascinating subject would, however, take us too far afield!

The concept of *expectation,* which we have discussed extensively in the previous pages, is of vital importance well beyond the theory of gambling games. In statistics, we sometimes call it the *expected value* or *mean.* Thus if we estimate the height of a distant building 10 times and get the estimates (in feet)

121, 120, 120, 121, 119, 120, 121, 119, 122, 120,

then we may compute the average of these figures, getting

$$\frac{121 + 120 + 120 + 121 + 119 + 120 + 121 + 119 + 122 + 120}{10}$$

= 120.3.

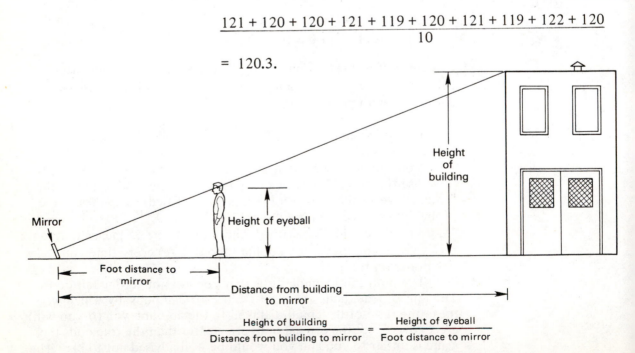

$$\frac{\text{Height of building}}{\text{Distance from building to mirror}} = \frac{\text{Height of eyeball}}{\text{Foot distance to mirror}}$$

On the other hand, taking the point of view of our recent discussion, we may say that there were four outcomes to our experiment (119, 120, 121, 122) and that the probabilities associated with these outcomes were $\frac{2}{10}, \frac{4}{10}, \frac{3}{10}, \frac{1}{10}$. Thus the expectation is given by

$$E = \left(\frac{2}{10} \times 119\right) + \left(\frac{4}{10} \times 120\right) + \left(\frac{3}{10} \times 121\right)$$
$$+ \left(\frac{1}{10} \times 122\right)$$
$$= 23.8 + 48 + 36.3 + 12.2$$
$$= 120.3.$$

This relation of average to expectation is actually quite general, and demonstrates the close connection of the probabilistic ideas we have been describing in this and the previous section with the familiar idea of average and with important notions of statistics. Outside the study of games, no particular significance attaches to the notion of 'fairness'—it would be silly to describe an experiment as 'fair' if the expected, or average, value of the quantity being measured is zero! However, in scientific work, we do sometimes 'renormalize' our measurements by centering them at the mean. This means that we record deviations from the mean (instead of the actual measurements themselves). The expected value of these deviations is then zero. In our example above, the deviations are

$$0.7, -0.3, -0.3, 0.7, -1.3, -0.3, 0.7, -1.3, 1.7, -0.3,$$

and you may readily verify (writing out the average or calculating E) that their mean or expectation is zero. Further discussion of averages will be found in the next section.

* In scientific experiments there is a method of calculating probabilities which is extremely important and which involves odds in a very essential way†. It is called *Bayes' principle,* or the *principle of inductive inference.* Suppose you formulate a hypothesis H (for example), that a coin is so biased that it is bound to show heads when tossed. Suppose further that you regard the only alternative as being that the coin is fair, and that you estimate the odds on hypothesis H to be q (say 1 to 9). You now carry out an experiment having a certain result R, namely, you toss the coin and it shows up heads. What are now the odds on your hypothesis? They are clearly now better

†The rest of this section if, of course, to be regarded as optional reading. It is recommended especially for those interested in the design of scientific and sociological experiments and the interpretation of their outcomes. It is likely to prove difficult for those readers who really need the rest of this chapter, but highly motivated students should make the effort. As we say in the introduction, you should not expect always to understand something the first time you meet it.

(bigger) than q, but how do we calculate them? The answer is given by the following formula which you can justify, if you want to go more deeply into the theory, by carefully applying the methods of conditional probability. The new odds, $\tilde{q}$, are related to the old odds q by the rule

Bayes' Formula

$$\tilde{q} = \frac{\text{probability of getting } R,\text{ given that hypothesis } H \text{ is true}}{\text{probability of getting } R,\text{ given that hypothesis } H \text{ is false}} \times q = F \times q.$$

In our case we compute $\tilde{q}$ as follows. If hypothesis H is true, that is, if the coin is certain to show up heads, then the probability of getting R is 1. If hypothesis H is false, that is, if the coin is true, then the probability of getting R is $\frac{1}{2}$. Thus our factor F is $1 \div \frac{1}{2}$. Since $2 \times \frac{1}{2} = 1$ we see that $1 \div \frac{1}{2} = 2$ so that $F = 2$. Thus $\tilde{q} = 2q = \frac{2}{9}$, if $q = \frac{1}{9}$. Indeed, we see that every time we perform the experiment and get heads, the odds go up by a factor of 2—it is for this reason that we compute F as a distinct quantity.

A real difficulty in this method is the estimate of the *prior probability, p,* or *prior odds, q.* In certain circumstances this can be done quite accurately, but often we are involved in little more than guess work. Very often, however, we know that the prior odds must be very small, so that even a surprising outcome to our experiment does not make the *posterior odds, $\tilde{q}$* into a big number. Coincidences do occur, and should not lead us to abandon well-established beliefs (in the validity of physical laws, for example), or to adopt strange new beliefs precipitately (such as extrasensory perception).

The calculation of the factor F converting the prior odds q into the posterior odds $\tilde{q}$ plainly requires skill in dividing fractions. This will be provided explicitly in the next chapter.

Exercise 2.7

These exercises will help you to become more familiar with the definitions connected with probability and odds. Use the information given in Problems 1 through 10 to complete the following table. Notice that the fractions that now appear are *not* reduced. If you adhere to this policy in filling out the rest of the table the final result will be more illuminating (and it will also save you some effort).

	I	II	III	IV	V	VI	VII
Problem number	Number of favorable outcomes	Number of un-favorable outcomes	Total number of outcomes	Probability of a favorable outcome	Probability of an unfavorable outcome	Odds on a favorable outcome	Odds on an unfavorable outcome
1.	2	4	6	$\frac{2}{6}$	$\frac{4}{6}$	$\frac{2}{4}$	$\frac{4}{2}$
2.		5					
3.							
4.							
5.							
6.			27				
7.			400				
8.			15 or (15 × 24)				
9.							
10.							

T.G.I.S. PARTY

1. An ordinary die is tossed. It is a favorable outcome if you throw either a 1 or a 2.

2. A day of the week is selected at random. It is a favorable selection if the name of the day selected begins with the letter *S*.

3. A digit is selected at random from the set designated by $\{0, 1, 2, 3, 4, 5, 6, 7, 8, 9\}$. It is a favorable selection if the digit selected is either even or a multiple of 3. (Remember that 0 is an even number.)

4.   A card is drawn at random from an ordinary bridge deck. A favorable draw is any face card (that is, a jack, a queen or a king).

5.   A sack contains 14 red poker chips, 7 blue poker chips and 15 white poker chips. A person is told to reach into the sack and draw out one poker chip. The person has succeeded if a red poker chip is drawn.

6.   A drawer is known to contain 9 red socks, 12 blue socks and 6 green socks. Suppose you enter the room in the dark and draw one sock from the drawer, and since you already have one red sock on before you enter the room, you will consider it a success only if you draw a red sock.

7.   A bag of money is known to contain dimes, having a total face value of $40.00. It is also known that 60 of the coins are solid silver and that the remaining coins are the less valuable 'sandwich' variety (containing mostly copper). You are allowed to reach into the bag and select one dime. Since silver is so much more valuable than copper you consider it a success if you draw a silver coin.

8.   Unmarked identical cans of fruit are displayed for sale in a large bin at the supermarket. They are offered at half price. The customer is told, by the grocer, that the bin contains 3 cases of pineapple, 4 cases of peaches, 6 cases of pears and 2 cases of sliced apples (each case contains 24 cans). The customer knows that his family will eat pineapple, pears or peaches, but not sliced apples. Therefore, if he selects one can, it will be a successful purchase if it is not sliced apples.

9.   A letter is selected at random from the alphabet. The selection is favorable if the letter chosen comes *before* the letter K in the alphabet.

*   10.   A number is selected at random from the set: $0, 1, 2, 3, \cdots, (D-1)$. The selection is favorable if the number chosen comes before the number N (which must be less than or equal to $D-1$). (Hint: If you have trouble here try some examples. Thus, you might look at what happens when $D = 9, N = 5$; or when $D = 7, N = 4$, etc.)

11. Carefully study the table for Problems 1 through 10 and observe the relationships between the numbers in various columns.

   (a)   For each problem what do you notice about the sum of the numbers in columns I and II?

   (b)   For each problem what do you notice about the sum of the numbers in columns IV and V?

   (c)   For each problem what do you notice about the product of the numbers in columns VI and VII?

12. Reconstruct the table on page 105, which was like the one on page 103 before some of its entries were obliterated.

13. Suppose you play a game in which you pay $3 to roll a pair of dice. You win $10 (and recover your stake) if you roll a 7 or an 11, otherwise you lose your stake.

   (a)   What is the probability of winning $10 on a single roll? (Ans: $\frac{8}{36} = \frac{2}{9}$)

   (b)   What is the probability of losing $3 on a single roll? (Ans: $\frac{28}{36} = \frac{7}{9}$)

   Suppose you played this game 100 times.

   (c)   About how much money would you expect to win on your winning rolls (altogether)? (Ans: $\$\frac{2}{9} \times 10 \times 100 = \$222.22$)

Problem number	I Number of favorable outcomes	II Number or un-favorable outcomes	III Total number of outcomes	IV Probability of a favorable outcome	V Probability of an unfavorable outcome	VI Odds on a favorable outcome	VII Odds on an unfavorable outcome
A.							$\frac{7}{3}$
B.				$\frac{1}{8}$			
C.		6	8				
D.	3	6					
E.					$\frac{7}{9}$		
F.						$\frac{3}{5}$	

    (d)   About how much money would you expect to lose on your losing rolls (altogether)? (Ans: $\$\frac{7}{9} \times 3 \times 100 = \$233.33$)

    (e)   After the 100 games would you expect to be ahead or 'in the hole,' and by how much? (Ans: In the hole by $11.11)

    (f)   What is your expectation in this game?

Suppose you played this game 1,000 times.

    (g)   About how much money would you expect to win on your winning rolls (altogether)?

    (h)   About how much money would you expect to lose on your losing rolls (altogether)?

    (i)   After the 1,000 games would you expect to be ahead or 'in the hole,' and by how much?

    (j)   What are the odds on winning this game? Assuming you continue to win $10, if you roll a 7 or an 11, what should you be required to pay each time you play for this to be a fair game?

14.   Suppose a roulette wheel has the numbers 0 through 36 and also a double zero (00), for a total of 38 slots. Neither the 0 nor the 00 are colored and the numbers 1 through 36 are colored alternately black and red as described in Example II. The betting is as described in Example II on the colored numbers only.

    (a)   What are the true odds for this game?

    (b)   If the official odds are 1 to 1, about how much will you lose, on the average, each time you play?

15.   An urn contains 24 balls that are known to be black, yellow or red. We conduct the experiment, 100 times, of taking a ball at random from the

urn recording its color, and returning it to the urn. Our experiments yield the following results:

a black ball was drawn 63 times;
a yellow ball was drawn 29 times;
a red ball was drawn 8 times.

(a)   How many balls of each color do you think are in the urn?

(b)   Assuming you are right, and that you lose $1 if you draw a black ball, win $2 if you draw a yellow ball, and win $3 if you draw a red ball, what is your expectation (each time you play the game, or conduct the experiment)?

*   16.   I know that, among my collection of 100 dice, there is one so weighted that it can only come down 3 or 5. I choose a die at random and want to decide, by throwing it, if it is the biased one. I throw it 6 times and get a 3 or 5 each time. What is the probability that I have chosen the biased die?

2.8 AVERAGES AND FRACTIONS

It is customary in many sports to try to represent the skill or success of a player by expressing the ratio of successful ventures to the total number of ventures as a fraction and calling this fraction the player's *average*. Thus, for example, in baseball, if the hitter scores 9 hits out of 30 times at bat, he or she is said to have an average of $\frac{9}{30}$. Hitter A is regarded as better than hitter B if A has a *better average,* that is, if the fraction representing A's ratio of successful hits is greater than the fraction representing B's ratio of successful hits.

This method of comparing players is fraught with difficulties and paradoxes—nevertheless, it is so well established that we will have to live with it. Let us take an example to show the real difficulty here. Suppose hitter A has had 40 hits out of 110 times at bat

in the first half of the season and 30 hits out of 40 times at bat in the second half of the season; while hitter B has had 10 hits out of 30 times at bat in the first half of the season and 20 hits out of 30 times at bat in the second half of the season. *Then A has a better average than B over the first half of the season and over the second half of the season, but not over the whole season*! To see this, we first compare the fractions $\frac{40}{110}$ and $\frac{10}{30}$. To do this we can reduce the fractions to $\frac{4}{11}$ and $\frac{1}{3}$, and use the common denominator 33 ($= 11 \times 3$) to replace the fractions by $\frac{12}{33}$ and $\frac{11}{33}$. (Alternatively, we can approximate the fractions by decimals, getting $.36 \cdots$ and $.33 \cdots$.) We next compare the fractions $\frac{30}{40}$ and $\frac{20}{30}$, or $\frac{3}{4}$ and $\frac{2}{3}$, and again use a common denominator (12) to show that $\frac{3}{4}$ is bigger than $\frac{2}{3}$ since $\frac{3}{4} = \frac{9}{12}$ and $\frac{2}{3} = \frac{8}{12}$. Or, using the decimal approximations we see that $\frac{3}{4}$ is approximately $.75$† and $\frac{2}{3}$ is approximately $.67$. Thus A certainly has a better average than B over the two halves of the season considered separately. But over the whole season A scored 70 hits out of 150, B scored 30 hits out of 60, so their averages over the entire season are $\frac{70}{150}$ and $\frac{30}{60}$, or $\frac{7}{15}$ and $\frac{1}{2}$. Now $\frac{7}{15} = \frac{14}{30}$, $\frac{1}{2} = \frac{15}{30}$, so B has a better average than A over the whole season!

How can this be? To understand the arithmetical processes described above, we should recall the distinction we made in Section 2.3 between *fractions* and *rational numbers*. A fraction is a symbol $\frac{N}{D}$, having a numerator N (or 'top') and a denominator D (or 'bottom') both of which are whole numbers ($D \neq 0$). Such fractions can be used to measure amounts, and equivalent fractions measure the same amount. For example, the equivalent fractions $\frac{2}{4}$ and $\frac{3}{6}$ represent the same amount of a given thing. A rational number is then an amount represented by a collection of equivalent fractions; we call it a number because it *can* be thought of as a number just as a decimal is a number. Now when we compare two fractions for size, asking which is the larger, we are *really* comparing the rational numbers they represent—indeed, in our procedure for making the comparison, we did replace the fractions representing the averages by equivalent fractions. But in combining the two fractions representing the average for the two halves of the season we actually worked with the fractions themselves. We introduced the operation (which we will write as $\vee$),

$$\frac{40}{110} \vee \frac{30}{40} = \frac{70}{150}.$$

(Note: $\frac{70}{150}$ is called the *mediant* of $\frac{40}{110}$ and $\frac{30}{40}$.) This is a perfectly good operation on (pairs of) fractions, but it is *not* an operation on

---

† $\frac{3}{4}$ is, of course, exactly .75. Anything approximates itself!

rational numbers! What do we mean by this? Well, the operation may be described in general as

$$\frac{A}{B} \vee \frac{C}{D} = \frac{A + C}{B + D} ,$$

and we don't get the same answer on the right if we replace $\frac{A}{B}$, say, by an equivalent fraction—the two answers on the right will not, in general, be equivalent. For example

$$\frac{1}{2} \vee \frac{1}{3} = \frac{2}{5} , \qquad \frac{2}{4} \vee \frac{1}{3} = \frac{3}{7} ,$$

and $\frac{2}{5}$ and $\frac{3}{7}$ are certainly not equivalent. So what we do with averages is a strange mixture. We *say* the average is a fraction but we *compare* averages as if they are rational numbers. We then *combine* averages by a rule which makes sense for fractions but is meaningless (that is, senseless) for rational numbers. It is no wonder that paradoxes result!

Many people confuse the operation $\vee$ with the addition of fractions. The rule we gave for adding fractions (Section 2.5), while harder to use than $\vee$, has the following two advantages:

(i)   it fits with the real-life applications described in Section 2.5 and Section 2.6;

(ii) it *is* an operation on rational numbers.

The operation $\vee$ is, of course, only one of many operations we may perform on fractions (especially if we do not insist that they should really be operations on rational numbers). Suppose we are set a target to achieve each day in our work (the same amount each day). Suppose we achieve $\frac{1}{2}$ of our target on Monday and $\frac{2}{3}$ of our target on Tuesday. What fraction of our target for Monday *and* Tuesday have we achieved? Plainly the answer is not $\frac{1}{2} + \frac{2}{3}$, although we may have the feeling we should add something to get the answer. A little thought shows that the answer is $\frac{1}{2} \times (\frac{1}{2} + \frac{2}{3})$. Thus we have introduced a new operation—a sort of averaging of fractions—by the rule

$$\frac{A}{B} \wedge \frac{C}{D} = \frac{1}{2} \times \left(\frac{A}{B} + \frac{C}{D} \right) .$$

This is a useful operation—and it *is* an operation on rational numbers, as you can see either by looking again at the problem which produced it or by looking at the rule itself, since adding and halving are both operations on numbers. However, it is *not* our rule for adding fractions—and it has the disadvantage of not being *associative,* that is, it is *not* true that

$$\left(\frac{A}{B} \wedge \frac{C}{D} \right) \wedge \frac{E}{F} = \frac{A}{B} \wedge \left(\frac{C}{D} \wedge \frac{E}{F} \right) .$$

Try this with your own choice of numbers A, B, C, D, E, F.

Not all averages are worked out by taking 'half of the sum' (that is, the midpoint). Suppose you bicycle from home to work at 15 mph and back from work to home at 10 mph. What is your average speed for the double journey? The answer is not $12\frac{1}{2}$ mph! For suppose the distance from home to work is 30 miles (that's a convenient number, not a desirable situation). Then it would take you 2 hours to go from home to work and 3 hours to go from work to home. It follows that it would take you 5 hours to complete the double journey of 60 miles. So your average speed would by 12 mph. The operation we have introduced here, to compute such average speeds, takes on the following form in the general case of two speeds of A mph and B mph:

(8.1) $$A \bigcirc B = \frac{2 \times A \times B}{A + B}.$$

In our example $A = 15$ and $B = 10$, so

$$A \bigcirc B = \frac{2 \times 15 \times 10}{15 + 10} = \frac{300}{25} = 12.$$

You are not expected to be able to deduce the general rule, but you may wish to try it with other convenient numbers. Properly interpreted we can use the operation $\bigcirc$ when A and B are themselves fractions—but this would again lead us to the study of division of fractions, and division is the topic of the next chapter.

You may be interested to know that the quantity $A \bigcirc B$ is called the *harmonic mean* of A and B, and is important in musical theory.

Exercise 2.8

1. In a double header, during the first game, Alice (A) made 1 hit in 3 times at bat and Betty (B) made 2 hits in 6 times at bat. During the second game they both made 2 hits in 3 times at bat. So their averages were the same in each of the two games. How did their averages compare if you take into account *both* games?

2. The following chart shows batting averages for Al (A), Bill (B), Charles (C) and Dave (D) during the first and second half of the season.

	A	B	C	D
First half	$\frac{20}{60}$	$\frac{10}{30}$	$\frac{10}{30}$	$\frac{40}{110}$
Second half	$\frac{20}{30}$	$\frac{20}{30}$	$\frac{40}{60}$	$\frac{30}{40}$

(a)   Find the overall average for each of the players.

(b)   How do the player's averages compare in the first half of the season?

(c)   How do their averages compare in the second half of the season?

(d)   How do their overall averages compare? (Hint: Write each of the averages as an equivalent fraction so that they all have the same denominator.

3.   Suppose that, instead of taking the average as is customary in judging baseball players, you measure their skill by 'hit advantage', where

hit advantage = (number of hits) − (number of misses).

(a)   Show that hit advantage is not subject to the strange defects of the traditional average.

(b)   Do you see any logical reason against replacing the average by the hit advantage?

(c)   Do you think there is any chance that the average will be replaced by the hit advantage?

*   4.   Show that, given two rational numbers $\frac{A}{B}$ and $\frac{C}{D}$, then

$$\frac{A}{B} > \frac{C}{D} \text{ if, and only if, } A \times D > B \times C.$$

(Notice that this rule would have simplified many of the calculations in this section and in some of these problems.)

5.   Suppose you jog from home to the park at 10 mph and jog home at 5 mph. Compute the average speed traveled during the double trip by finding the total time it took and dividing by the total distance traveled if you assume that the distance between your home and the park is

(a)   20 miles.

(b)   40 miles.

(c)   Find the average speed traveled on the double journey by using formula (8.1) where $A = 10$ and $B = 5$.

(d)   What is the average speed if the speed in each direction is 7 mph? Do you recognize a general principle here?

(e)   Is the average speed greater if you jog from A to B at 10 mph and back at 5 mph, or if you jog in both directions at 7 mph? In which case is the *average of the speeds* greater?

6.   (Harder) Show that if $A \bigcirc B$ is the harmonic mean of A and B, then $\frac{1}{A \bigcirc B}$ is the average (arithmetic mean) of $\frac{1}{A}$ and $\frac{1}{B}$.

CHAPTER 3

Division

Division presents problems which are not found in the other three arithmetical operations. Where division of whole numbers by whole numbers is involved, the correct answer must depend on the context. The division of fractions, which came up naturally in Section 2.7, is also discussed.

3.1 DANGEROUS DIVISION†

We have already examined several real-life situations where the necessity to divide either whole numbers or fractions occurs quite naturally. And, although it may be of less general interest it should be recognized that division does occur in many other contexts. For example, at the end of Chapter 2 the formula used to compute the average speed, C mph, for the double journey if we go from X to Y at A mph and back from Y to X at B mph is

$$C = \frac{2 \times A \times B}{A + B}.$$

This expression naturally requires us to carry out the division of $2 \times A \times B$ by $A + B$. This application shows that if you're interested in speed you must know how to add, multiply and divide, and will almost certainly need to be able to carry out these arithmetical operations with fractions and decimals as well as whole numbers.

Our point is simply this: intelligent people who wish to solve certain real-life problems are ultimately forced to gain an under-

†Much of the material of this section and the next appeared in an article by one of us (PH) in California MathematiCs 4 (1979), 15–21.

standing of the division concept. Fortunately it is not a concept of insurmountable difficulty (despite popular opinion!), but it will definitely be easier for you to use the concept if you look at it from the right point of view and examine enough examples to become alert to the generalities and the possible difficulties involved. It is certainly true that it is not a trivial concept and there are some hazards. That is why we have deferred this discussion until now, and it is also why we begin our treatment of the subject with the warning that DIVISION IS DIFFERENT!

There are various subtle respects in which division *is* different from the other three arithmetic operations. For example, when you add, subtract or multiply whole numbers you begin the actual computation by dealing with the digits on the right (the units). But, with division you must begin the computation by dealing with the digits on the *left*. By contrast with addition, subtraction and multiplication, there is no dependable rule that determines the units digit in the answer by using only the units digits in the two numbers in the original problem. Recall, for example, that in the problem $34 \div 2$ the last digit in the answer is 7; whereas the last digit of the answer in the problem $24 \div 2$ is 2. Without taking into account the tens digit to the left of 4 in the dividend in these divisions by 2 there is no way to know whether 7 or 2 should be the units digit in the answer. If we were dividing by some other number we might well need to know the hundreds digit, thousands digit, . . ., in the dividend. Thus it is not merely the particular division algorithm used that forces us to 'start on the left'.

Another difference is that we have many ways of writing a division problem and each of those ways seem to carry a different psychological message. Thus, we have

$$245 \div 7, \quad 7\,\overline{)245}, \quad \text{and} \quad \frac{245}{7}.$$

In the first expression we know we are going to be expected to do the division, and this way of writing also implies that we will be required to complete the expression by making a mathematical statement: $245 \div 7 = 35$. In the second notation there is no need to make a statement; indeed, it would be thought perverse of us to do so. Instead we write, rather cryptically,

$$7 \overline{)245} \quad \begin{array}{r} 35 \\ \end{array}$$

$$\begin{array}{r} 35 \\ 7 \overline{)245} \end{array},$$

perhaps adorning this mystic symbol with certain marks (evidential relics) of our actual working. The third notation is the least informative, in terms of what is expected. Many people would regard themselves as fully entitled to do nothing with the expression, while others might feel a compulsion to divide or to write it as a mixed number. You would have no basis for knowing which view is correct unless you knew where the expression came from, and what you wanted to do with it later. This question of where the division problem comes from will prove to be the most important one in understanding division.

Our first concern here, however, is with the fundamental, and very disturbing, respect in which division stands alone among the primary arithmetical operations on the set of whole numbers. It is a simple, easily observed, fact that while division is, mathematically, the inverse of multiplication, we are often, at a very early stage of our mathematical education, called upon to divide where no multiplication has, even in principle, taken place! For example, if my annual salary is $10,000, I may be paid in monthly installments. The amount of each installment is determined by the solution of the division problem $10,000 \div 12$. Yet there is no whole number (of dollars, or even of cents) which solves this problem (that is, there is no whole number c such that $12 \times c = 10,000$ or even such that $12 \times c = 1,000,000$). Other familiar situations give rise to comparable difficulties. We must divide: yet, strictly speaking, we *cannot*. Such situations we call cases of *dangerous division*.

There is no algorithmic (routine, or mechanical) procedure for handling dangerous division. Here is a sharp contrast with addition, subtraction and multiplication. Addition and multiplication are never dangerous exercises (though, as you might recall, they may be appallingly tedious) and they can always be done routinely. Subtraction problems occasionally have a small element of danger—it may be necessary to ask if, in the given context, a negative answer makes sense—but most subtraction problems are essentially routine computations. Division is only routine where it is genuinely the inverse of multiplication, and this is rare in 'real life.'

We are thus brought face to face with the heart of the real difficulty surrounding division: *How do you interpret the outcome of the formal division process so that you can respond appropriately*

to the original 'real-life' problem? Unfortunately many texts (spuriously and misleadingly) locate the difficulty elsewhere. First they foist on the suffering student a horrendous, cumbersome and almost entirely useless algorithm for the formal process—that's the thing called 'long division.' Then they suggest increasingly sophisticated ways of presenting the 'solution'—quotients and remainders, decimals, fractions—and these representations are chosen with total disregard for their meaning in the original problem. Indeed, usually no 'original problem' is even mentioned.

We begin the next section by poking some good-natured fun at the traditional way in which division is presented. In order that our readers may join wholeheartedly in the fun, we think it might be helpful for us to enliven the next section by reviewing the standard solutions of division problems. We do so not in order to help our readers to achieve mastery of the horrible algorithms against which they have almost certainly already rebelled in their chuldhood—we set no store by such mastery as a real-life skill—the calculator has been invented! We do so just in order that our readers may know what we are talking about when we describe that absurd distortion of an important piece of mathematics which is (mis)named 'division' in a traditional elementary mathematics program.

We are here only concerned with dividing one whole number by another, and we will concentrate on recalling the *meaning* of the three types of solutions—quotients and remainders, decimals, fractions—and will not describe in detail *how* the solution is obtained. We will, however, show by examples how the solutions may be justified. Were we to describe the algorithm usually inflicted on our suffering children for obtaining solutions of division problems, we would probably achieve all over again the effect which first acquaintance with that algorithm produced—healthy aversion to the ugly and superfluous!

We will assume in what follows that the reader understands the division algorithm for whole numbers and could carry it out if forced to do so. Any reader not entirely comfortable with the process of dividing one whole number by another, getting a quotient and remainder, should consult the relevant part of the Appendix at this point.

Exercise 3.1

1. Look in some elementary mathematics textbooks and record all of the different ways you can find for writing division problems.

2. Give three examples of real-life situations where you must divide, but strictly speaking you *cannot*.

3. Give a probable 'solution' for each of the examples you cited in problem 2.

3.2 THE TRADITIONAL PROGRESSION (ANXIETY REVISITED)

We will now sketch the traditional progression (of learning division) and point out some of the common offensive, and totally unnecessary, practices. Perhaps this will give you a better understanding about why so many people feel uncomfortable about division. And since our object is to *reduce* your anxiety about division we will then show how to avoid these pitfalls. To do this we will stress understanding what the division of one number by another really *means* in a purely mathematical context. Then we will discuss how it should be used, and interpreted, in various real-life situations. First then, for how division is usually presented—an example of 'how not to do it'!

Division (of whole numbers, by whole numbers) is usually presented as a well-defined arithmetical operation. For example, 12 divided by 3 is defined as the whole number c such that $12 = 3 \times c$. (Here c must be 4, so we say '12 divided by 3 is 4'.) The problems are then constructed so that the way the answers are obtained appears to become more complicated each time the student encounters the concept. Thus, initially, the student is only asked to divide a by b if there exists a whole number c such that $a = b \times c$† (for example, divide 6 by 2, or divide 20 by 4, etc.). If the numbers a, b, c are small enough that $a = b \times c$ may be regarded as a 'number fact' the procedure is this: if you want to divide a by b, simply consult your memory bank to identify the number c you multiply by b to get the answer a, and then announce that a divided by b is c ($a \div b = c$). If the numbers a and b are large, and chosen with regard to the fact that we use a base ten system of numeration, then one may use a pleasant and efficient algorithm to obtain the answer. Thus, for example, the problems $10,000 \div 10$, or $800 \div 5$ lead to quite painless, indeed positively useful and informative, experiences for the student. However, the honeymoon is quickly over; and before the student has a chance to escape he or she is now trapped in the 'long division algorithm', a procedure of vanishing utility (in view of the advent of the hand calculator) and unsurpassed in its capacity to induce boredom and distaste in those obliged to practice it.

At a certain stage it is admitted that, given a and b, there may exist no whole number c such that $a = b \times c$. The students really knew this all along and they generally know how to handle the real-life situations that produce such problems too. What the students are hoping is that the teacher won't mention it, because they fear

---

†In case you have trouble remembering that $a \div b = c$ means $a = b \times c$, think how you set out the division problem. To find $a \div b$ you write $b \overline{)a}$ and, once the division is done, c appears to complete the picture,

$$b \overline{)a}^{\,c} .$$

Thus the routine check is to verify that $b \times c = a$.

that the mathematical model is likely to be more obscure than the real-life situation they understand. Their feelings are soon justified as the concept of 'remainder' is presented and they find that the expected solution to the problem $17 \div 4$ is 4R1. This now introduces an element of mystery into division which was not present in the other three arithmetical operations—no strange new symbols had to be introduced to solve problems involving addition, subtraction or multiplication.† Moreover, if we have \$17 and wish to purchase some ties costing \$4 each, then, in the solution 4R1, the '4' stands for ties and the '1' stands for dollars! But it turns out that this notation not only forces them to accept a new symbol; the new symbol itself quickly offends their sense of reason! For instance, students will already know that since $(12 + 7)$ and $(8 + 11)$ are both 19, they are entitled to write $12 + 7 = 8 + 11$; indeed, such statements are not only true but they are valuable. However, students would be well advised *not* to write $17 \div 4 = 161 \div 40$; even though both divisions have the solution 4R1. This difficulty is traditionally overcome by the standard pedagogical device of simply not mentioning it! The effect, of course, is to reinforce the students' feeling that numbers in mathematics classes often don't relate to the real world they live in, and that mathematics is unnecessarily complicated.

In more advanced classes the solution is presented as a decimal, up to any required degree of accuracy (generally requiring successively one more digit to the right of the decimal point in each of the sixth, seventh, and eighth grades). Thus we see that $17 \div 4 = 4R1$ is now replaced by such sophisticated statements as $17 \div 4 = 4.25$. There is some difficulty if we merely require accuracy to the nearest tenth (because $\frac{5}{100}$ is *exactly* in the center of the interval between $\frac{0}{100}$ and $\frac{10}{100}$), but this is usually abolished by some arbitrary convention. There is a further difficulty because some problems (like $23 \div 7$) do not have a finite decimal answer. Even if we are asked for accuracy to the nearest hundredth, we still should not write $23 \div 7 = 3.29$. We might try adopting the notation $23 \div 7 = 3.29 \cdots$, but those dots tend to suggest the real answer is bigger than 3.29. Moreover, they have the magical property that even if they are filled in, they still reappear! Thus we cannot write $23 \div 7 = 3.28571$, but we can write $23 \div 7 = 3.28571 \cdots$!

We suspect most students give up trying to *understand* division at this stage, but nevertheless—or perhaps because of this attitude of resignation—they become reasonably good at 'getting the right answer' But now even their efficiency is likely to be disturbed by a new development which occurs when the study of fractions is prosecuted with due fervor by the teacher. Here division is disguised in a process

---

†Of course, subtraction may lead *later* to the introduction of negative numbers.

called replacing an 'improper fraction' by a 'mixed number'. The student is expected to learn that $17 \div 4$ and $4\frac{1}{4}$ are the same thing because $\frac{17}{4} = 4\frac{1}{4}$. What is remarkable is that so many students survive the cultural shock of this intellectual assault! '$17 \div 4$' is, for them, an order to do a certain piece of arithmetic; whereas $\frac{17}{4}$ is a fraction. Suddenly, both $17 \div 4$ and $\frac{17}{4}$ become numbers—and they become the same number! If the student has been paying attention, mystification should now be complete.

We do not claim that every student is exposed to every one of these obfuscations. But our experience with students has shown us how typical this pedagogy is. Our point is that the appropriate way to handle a division problem should depend on the real context of the problem and not the grade level of the student.

Now for some common sense! What we will show in the rest of this section is that, except in the original situation where there is a whole number c such that $a = b \times c$, the instruction to solve $a \div b$ is incomplete without a statement of the context; and that the 'solution' to a division problem depends on knowing that context. Here we use the word 'solution' in the sense of the answer to a real-life question *requiring* us to carry out a division. The word 'solution' is sometimes used to mean the process of carrying out the division, sometimes to mean the quotient itself. We mean something different from both of these—and more fundamental. It is far more difficult to decide what degree of approximation is appropriate to the answer to the real question than to carry out some dreary algorithm—the former is a strictly human activity while the latter is usually best left to a machine (though we describe a 'hand algorithm' in the Appendix).

Since for any numbers a and b ($\neq 0$) you can obtain a decimal approximation for $a \div b$ on a hand calculator you usually don't need to know the more sophisticated methods of division mentioned above (but you do need to recognize wildly incorrect answers—so that you will know when you have pushed the wrong key on your calculator). It would be far wiser for you to concentrate on learning how to estimate approximately what the answer should be, and to learn more than one way to obtain the answer on your calculator (as a double check). Having done this your energies can best be spent then by learning how to adapt division to a wide variety of concrete situations.

In order to see how diverse those applications can be, study the following seven problems. For each of them the appropriate mathematical model is $113 \div 6$, yet each has a different answer! You should be able to think of other examples to add to this list.

Problem 1: A consignment of 113 guinea pigs arrives at the Darling's house, with instructions from their Uncle Jack that they are to be distributed equally among the 6 Darling children. How many does each child get?

Answer: Uncle Jack's instructions cannot be carried out. He is asking the impossible. Of course, the children might modify Uncle Jack's instructions, thus creating a different situation.

Problem 2: It is a state regulation that, on any official outing, there must be one adult for every 6 children. If 113 children go on an outing, how many adults must accompany them.

Answer: 19

Problem 3: The round-trip fare from my home to Rochester Bus Station is $6.00. How many such trips can I make if I have saved $113?

Answer: 18

Problem 4: I have 113 dollar bills which I propose to distribute among 6 people so that each gets an equal amount. What should I do to give as much as possible to each person?

Answer: Give each person 18 and retain 5. (We might write this 18R5.)

Problem 5: I have $113 which I propose to distribute among 6 people so that each gets an equal amount. What should I do to give as much as possible to each person?

Answer: Give each person $18.83 and retain 2 cents.

Problem 6: Cycling round a 6-mile track, my odometer records that I have covered 113 miles. How many times have I gone around the circuit?

Answer: $18\frac{5}{6}$

Problem 7: My model train covers 113 feet when it goes 6 times round the track. How long is the track?

Answer: 18 feet, 10 inches

The main point about the above problems is not that the various solutions involve different units. After all, this is also the nature of the other arithmetical operations. Moreover, it is as true of multiplication as of division that the units in the solution depend in a fairly sophisticated way on the units given in the problem. This is in marked contrast to addition and subtraction. For, in the latter two cases the given numbers must be in the same units for their sum or difference to be meaningful, and then, of course, the answer is in those same units.†

No, the main point is a mathematical one; and since arithmetical training is, traditionally, not accompanied by any education in mathematics, it is apt to be entirely overlooked. The point is that the whole numbers are not closed under division (which simply means that if you wish to divide one whole number by another whole number you might not be able to find a suitable answer among the whole numbers). You have encountered this before, since the whole numbers are also not closed under subtraction; but, in the case of subtraction, you were probably only called upon to subtract b from a when a is a number obtained from b by adding some whole number c. Thus subtraction genuinely presents itself as inverse to addition. However, although division is initially defined, and should be presented, as inverse to multiplication, problem situations almost immediately arise, in a natural way, in which we have to 'divide' a by b where there is no number (that is, whole number) c such that $a = b \times c$. To handle these situations we must adapt the division process and the adaptation, as we have seen in the examples, must depend on the problem context.

Some people claim you can get around the difficulty by passing from the whole numbers to the rational numbers. This is simply not so! In the first place, we are only dividing by whole numbers—division by rational numbers is a very different matter. But, even more important, the rational number $\frac{a}{b}$ may not be the correct answer to a problem for which the arithmetical model requires us to divide a by b. As we indicated in Problem 1, no sane person would think that 113 guinea pigs should be shared among 6 children by giving each child $18\frac{5}{6}$ guinea pigs!

Where division may not be carried out exactly‡ when dividing whole numbers, the division process is a *means* to the solution of a

---

†It is not true, as certain elementary texts suggest, that "6 pigs and 4 cows equal 10 animals". Arithmetic is not a process whereby you lose valuable information! Moreover, if you believe (correctly!) that if $a = c$ and $b = c$, then $a = b$, you will be troubled by the inference that 6 pigs and 4 cows equal 7 dogs and 3 cats! Of course, in some situations, some people (especially children) would wish to note that they have 10 animals if they have 6 pigs and 4 cows; but they would not use the notion of equality.

‡Many texts talk of the number b dividing 'evenly' into the number a. We enter a plea for the use of the term 'exactly.' It seems to us deliberately confusing to say that 3 divides *evenly* into 21, when the quotient is the *odd* number 7. Personally, we don't even like to talk of dividing *into*, since this, too, generates confusion. Remember, if you do meet the phrase, that 'dividing b into a' means the same as 'dividing a by b.'

real problem, and does not determine the solution. The choice of solution requires an understanding of the context and real-life alternatives (is it possible to get change for a dollar bill?) and is *not* simply dependent on your ability to execute sophisticated division algorithms. What you really need to learn is how to interpret the answer you got on your hand calculator, in terms of the problem that generated the calculation. The problems at the end of this section will give you some practice using your hand-calculator and looking at this aspect of solving problems that involve division of whole numbers.

The hand-calculator is useful for obtaining the decimal approximation for any given division problem involving whole numbers. But in order to use the result you must understand how to interpret the answer and frequently it is necessary to be able to use the result to obtain a slightly different formulation of the computation. For example, suppose you want to know how many silver dollars you will have left over after distributing 27 dollars among 4 people so that each person gets an equal maximum number of dollars. You can compute $27 \div 4$ on a hand-calculator to obtain 6.75; but this does not tell you the number of dollars you will have left over. However, since the whole number part of the answer is 6 it does tell you that each of the four people will get 6 dollars. You can deduce that 4×6, or 24, of the dollars were used in the distribution process and consequently $27 - 24$, or 3, dollars are left over. This example was intentionally constructed with small numbers so that it would be easy to understand (and you probably would have had no trouble working it without the aid of a calculator). What is important is the concept, and where the numbers are large you will probably really want to use a calculator. The idea, stated in general terms is this:

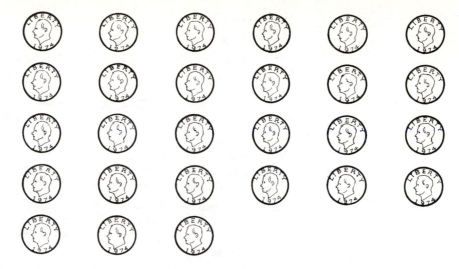

Remember that when you divide *a* by *b,* you get a quotient *q* and a remainder *r,* less than *b,* and that $a = (b \times q) + r.$

When you use a hand calculator to compute $a \div b$ the part of the answer to the *left* of the decimal point is *q.* You can then determine the value of *r* by subtracting $(b \times q)$ from *a.* This will enable you to write *a* as

$$a = (b \times q) + r.$$

Thus, in the above numerical example, we could write $27 = (4 \times 6) + 3.$ The answer to our actual question (how many silver dollars are left over?) is then 3.

Carrying out a few examples with small numbers should convince you that this process is always valid. Of course even if the numbers are large you can still check the result by carrying out first the multiplication and then the addition (perhaps using the calculator) on the right hand side to verify that the expression equals the number on the left.

We now give an example using larger numbers. Suppose you are using your hand-calculator to determine the *remainder* in the division problem $879 \div 56.$ You first carry out the computation $879 \div 56$ on your hand calculator, obtaining an answer of 15.69642857. This implies that 879 is equal to 56×15 plus a remainder. The remainder must be $879 - (56 \times 15),$ which is $879 - 840,$ or 39. The whole result may be expressed in the form $879 = (56 \times 15) + 39.$

Exercise 3.2

In Problems 1 through 12, using a calculator, if necessary,
(a)   find the decimal approximation for the indicated division,
(b)   write the result of the division $a \div b$ in the form $a = (b \times q) + r.$

1.	$42 \div 15$		**7.**	$970 \div 281$
2.	$37 \div 8$		**8.**	$76423 \div 1862$
3.	$472 \div 30$		**9.**	$4444 \div 1112$
4.	$560 \div 21$		**10.**	$10001 \div 101$
5.	$4972 \div 320$		**11.**	$6392 \div 68$
6.	$7866 \div 463$		**12.**	$400000 \div 1234$

Problems 13 through 24 involve the division of certain whole numbers. Determining the appropriate answer, however, requires an understanding of the context that gave rise to the calculation. You may find some of the results in Problems 1 through 12 helpful in finding appropriate answers to these problems.

13. Forty-two visiting orchestra members are to be housed in the homes of 15 local music patrons. If the orchestra members are to be distributed among the patrons as equally as possible, how many patrons will have three guests?

14. How many 8¢ stamps can be purchased with 37¢?

15. If the state law dictates that no class may have more than 30 students, what is the smallest number of classes that can be arranged in a school with an enrollment of 472 students?

16. When 560 new employees arrive for work at a new plant the personnel director discovers that her instructions are to send an equal number of new employees to each of the company's 21 different departments. What should she do?

17. If you have a bank balance of $4972 and want to buy IBM common stock, which is selling for $320 per share (including the broker's commission), how many shares can you buy? How much money, if any, will you have left over?

18. In June $7866 is to be disbursed among 463 employees as a month-end dividend for their share of the company's profit-sharing plan. The company has a policy to round off to the nearest cent; so that they either keep the excess in the fund until next month, or debit the fund with the additional amount required for the 'rounding up'.

    (a)   How much will each employee receive at the end of June?

    (b)   As a result of the June disbursement how much will July's account be credited, or debited?

19. The proprietor of a yard goods store has 970 feet of material with a pattern that repeats every 3 feet in such a way that an attractive table cloth can be made from a 3-foot length. Use an adaptation of the result of Problem 7(b) above to determine how many table cloths can be cut from the piece of material. (Hint: Since $970 = (281 \times 3) + 127$ we know that after 281 tablecloths are cut there will be 127 feet of material left over.)

20. If an import tax is collected of $1.00 per 1862 feet of electrical wire, and if $1.00 is collected on any length of wire shorter than the specified 1862 feet, how much tax will be levied on 76423 feet of wire?

21. A salesman driving his assigned 1112-mile circuit reports that as of January 31 he has traveled 4444 miles since January 1. How many times has he traveled around the circuit this year?

22. The proprietor of a nursing home has in her purse a hundred dollar bill and a penny. She discovers that the local market is having a 1¢ sale on the shampoo used at the nursing home; that is, if you pay cash for one bottle priced at a dollar you can purchase the second bottle for 1¢ (and this includes the tax). What is the maximum number of bottles of shampoo she can buy and how much money will she have left over?

23. An amusement park ride covers 6392 feet during one complete ride. If each ride has 68 cycles how long is one cycle?

24. A new stadium is planned to seat 400,000 people in individual seats. The seats are shipped to the construction company in trucks having a capacity of 1234 seats per truckload. How many truck loads will be required to deliver all of the seats?

*25. The problem discussed on page 120 may also be computed using another method as follows. Since $27 \div 4 = 6.75$ it follows from the definition of division that

$$27 = (6.75) \times 4,$$

and thus

$$27 = (6 + .75) \times 4,$$

or

$$27 = (6 \times 4) + (.75 \times 4).$$

Now, if we carry out the multiplication of the last two numbers on the right we obtain

$$27 = (6 \times 4) + 3.$$

This means, in the general formulation on page 121, that an alternative method for finding r would be to multiply b by the part of the answer appearing to the right of the decimal point in the original calculation of $a \div b$.

You may, however, encounter a small problem using this method if the original computation of $a \div b$ involves an answer having more digits to the right of the decimal place than the display on your calculator can show. For example, computing $23 \div 7$ on a hand calculator (with a display of 8 digits) gives 3.2857143, but when you compute $7 \times .2857143$ you will obtain 2.0000001. Of course, this is not too serious since you would certainly then guess that r must really be 2 and you could easily verify that

$$23 = (7 \times 3) + 2.$$

What really happened was that the calculator 'rounded off' the last digit in the display and the decimal *shown* was a little too big. We emphasize the fact that the number shown may not be what the calculator carries internally. Most calculators are accurate to many more decimal places than the number of digits they display. Consequently it is generally advantageous to perform your calculations so that you don't have to re-enter decimal numbers that have more digits than your calculator can display. In the above example you could have first divided 23 by 7, obtaining the displayed number 3.2857143. If you subtracted 3 from the displayed number and then multiplied by 7 you would have obtained the expected answer of 2.†

Work problems 8, 9, 10, 11 and 12 using this method.

3.3 DIVISION OF DECIMALS

The division of decimals occurs in many of the real-life contexts in which division of whole numbers occurs, so that much that was said in the previous sections of this chapter is again relevant here. Let us give two examples of situations in which we may be called upon to divide by decimals and then discuss how actually to carry out the arithmetic—or how to check that our hand calculator has given the right answer.

Example 1   We are told that a rectangular piece of land has an area of 50 square meters. We measure its length and find the length to be 12 meters 65 centimeters, or 12.65 meters. What is its width? The answer, in meters, is obtained by dividing 50 by 12.65.

Example 2   On our holidays we see a nice tablecloth in a store, priced at $26.40. We decide to buy it and find ourselves charged $28.25. We realize that the extra must be due to the sales tax. What is the

† You might obtain on your calculator an answer of 1.9999999—and this you should interpret as 2 (see Problem 6 in Section 2.4).

rate of sales tax in the state in which we are taking our vacation?†
The difference between the amount charged and the published price
is $1.85, so the rate of tax, *as a percentage,* is given by dividing 1.85
by 26.40 and then multiplying by 100.

In both these examples the number we are dividing by—called
the *divisor*—is a decimal, not a whole number. If we actually have to
divide, by hand, a number N by a *whole number M*, then no real
difficulty arises from the fact that N is a decimal rather than a whole
number. What we do is first estimate, rather crudely, what the
answer should be; then we carry out the division ignoring the decimals; and finally we place the decimal point where our rough estimate
dictates that it must go. And, in any case, if there is doubt about the
answer, we can *check* our result by multiplying it by the divisor to
make certain that product gives the original number.

Suppose, for example, we are dividing 21.72 by 6. We first
notice that if the problem had been 21 ÷ 6 our answer would have
been bigger than 3 (but less than 4). Next we carry out whatever
algorithm we're happy with to divide 2172 by 6. Thus we may write

$$\begin{array}{r} 362 \\ 6\overline{)2172}\,. \end{array}$$

And we now know, because of our original estimate, that the decimal
point must follow the digit 3 in our answer, and that the true
answer, or *quotient,* is 3.62. If we wished to check this answer we
could multiply 6 by 3.62, obtaining 21.72, which would verify that
our division is correct.

---

†It is a practice confined, as far as we know, to the United States to advertise a price and
then add the tax to one's bill, rather than to include the tax in the published price.

You might note that if we now inserted the decimal points in the appropriate place for 2172 and 362 the computation would appear as

$$\frac{3.62}{6 \overline{)21.72}}$$

with the decimal points aligned vertically.

To place this problem in a more concrete setting may convince you of the reasonableness of the result. For instance, imagine that you are dividing $21.72 equally among 6 people, but that you do the calculation in cents instead of dollars.

Of course the usual difficulties of 'round-off' occur in such problems, just as they did in division of whole numbers by whole numbers in the previous section. We avoided this difficulty in our example by choosing our numbers carefully, but in real-life they have to be faced. We face them now in tackling Examples 1 and 2.

Example 1
(again!)

We have to divide 50 by 12.65. One possible approach is to *make the divisor a whole number.* Had our problem been 'If we have $50.00 and want to buy as many ping-pong paddles as we can, each costing $12.65', we would have been faced with the division 50 ÷ 12.65, but we could work in cents, then the division would have been 5000 ÷ 1265. This technique always works. If we must divide the decimal N by the decimal M, we make the decimal M into a whole number by multiplying it by a suitable power of 10 and then we multiply N by the same power of 10. We are left with the problem of dividing a decimal by a whole number. In this case we must divide 5000 by 1265 and we find the answer to be 3.95, to two places of decimals, so that the width of the plot of land is 3.95 meters, to the nearest centimeter.†

An alternative strategy—and, in some respects, a better one—is to first do a very approximate division. We argue that 50 ÷ 12.65 cannot be very different from 50 ÷ 12, so the answer is roughly 4. We now carry out a division algorithm *totally ignoring decimal points.* That is, we divide 50 by 1265 but pay no attention whatsoever to decimal points, just concentrating on getting the digits right (when doing this you can affix as many zeros as you like to the right end of 50). We get a provisional answer of 3952 · · ·, and, knowing the actual answer is roughly 4, we can put the decimal point, with complete confidence, between the 3 and the 9. This strategy is a good one if we have a hand calculator available, as our approximate answer serves as a check that we have used the calculator properly.

---

†Here, again, we see the importance of looking at the nature of the problem that generated the calculation. This same calculation would tell us that the answer to the related 'ping-pong paddle' problem is 3.

Example 2
(again!) Here we must divide 1.85 by 26.40 and then multiply by 100, to get the sales tax expressed as a percentage. Of course, we are perfectly free *first* to multiply 1.85 by 100 and then divide by 26.40. If we do this—and we recommend this move!—we are faced with the division $185 \div 26.40$. The first method described in Example 1 above now yields $1850 \div 264$. You may wish to carry on from there, but we prefer to carry out, explicitly, the steps of the second method. We have the division problem $185 \div 26.40$. This is approximately (say) $185 \div 20$, or about 9 (notice how 'rough' these steps can be!). Now we carry out the division $185 \div 264$ accurately, but ignoring decimal points, getting $70075 \cdots$. The computation might look like this.

$$
\begin{array}{r}
70075 \\
264\,)\overline{18500000} \\
-1848 \\
\hline
2000 \\
-1848 \\
\hline
1520 \\
-1320 \\
\hline
200
\end{array}
$$

etc.

Finally we refer back to the original question and appeal to common sense. We are never going to be charged an amount expressed more accurately than in dollars and cents; and sales taxes are not going to be excessively complicated amounts like 7.0075% (although, at one time, some cities in California had a sales tax of 6.25%). The sales tax in this state is seen to be 7%—and we could, indeed, have stopped the division after getting that first 7 in our answer, since the remainder was so small.

Two features of these examples deserve a remark. The more important is the observation, stressed in the previous sections, that the accuracy with which we present the solution is dictated by the actual problem being solved. In Example 1, it was natural to give the width in meters and centimeters, since the length was given in these units. In Example 2 we allowed our common sense to tell us what the answer must be—the 'inaccuracy' lay in the cost of the tablecloth, not in our calculation of the tax rate.

The second remark refers to a feature of Example 2; we had to divide by 26.40 and we could just as well have dropped the '0' and divided by 26.4. However, in *certain* contexts, it would be important to remember that the divisor was given as 26.40, rather than 26.4, in order to decide the accuracy with which to present the quotient. Imagine, for example, that, in Example 1 the length had been 12 meters 60 centimeters, instead of 12 meters 65 centimeters.

There is really nothing more to be said about division of decimals; the reader may however, need practice in recognizing 'division situations', and in executing divisions correctly. We provide exercises to aid in the development of increased efficiency.

Remark
Suppose you are faced with a computation of the form $6 \times (8 \div 3)$ and you have some vague recollection that it is possible to get the correct answer by computing instead $(6 \times 8) \div 3$—but you are not certain. The general result you are questioning is whether $(A \times B) \div C$ is always equal to $A \times (B \div C)$. Without studying a formal proof you can gain some reasonable assurance that this is valid if you try several specific cases and find that they all work. For example, if we carefully select our numbers we can look at several examples where both expressions are easy to compute. Thus,

$$(3 \times 6) \div 2 \text{ is equal to} \quad 3 \times (6 \div 2),$$

$$(4 \times 12) \div 6 \text{ is equal to} \quad 4 \times (12 \div 6),$$

and

$$(10 \times 20) \div 4 \text{ is equal to} \quad 10 \times (20 \div 4).$$

We conclude that the rule is *probably* valid. (This does not constitute a proof, but it does give you quick access to forgotten formulas.) Notice that we verified the formula when both expressions were easy to compute. Of course, the formula is *useful* precisely when one expression is much easier to compute than the other, as in our original example of $6 \times (8 \div 3)$. By computing $(6 \times 8) \div 3$, we avoid fractions.

What happens if the formula you want to use is *not* true? Well, suppose as an example, you are not certain whether or not $(34 \times 98) + 2$ is the same as $34 \times (98 + 2)$. (You have some interest in the problem because the second expression would certainly be easier to

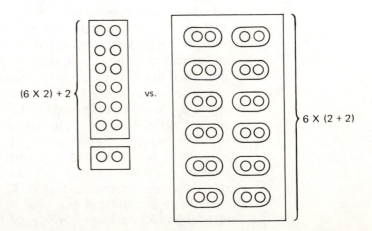

$(6 \times 2) + 2$ vs. $6 \times (2 + 2)$

compute.) This when stated as a general question appears thus: Is $(A \times B) + C$ always equal to $A \times (B + C)$? Having stated the problem in general, you can then look at some examples with smaller numbers to see if this would be valid. Letting $A = 6$, $B = 2$ and $C = 2$, we obtain the two expressions $(6 \times 2) + 2$, or 14, and $6 \times (2 + 2)$, which is 24. Since these two expressions are *not* the same the rule does not hold. So, in order to get the correct answer in the original problem you must first compute (43×98) and then add 2.

Notice that one failure, as above, is sufficient to show that a rule is wrong. However, to show that a rule is right it is important to look at *several* cases, to feel certain you have not accidentally chosen the only numbers for which the statement is true. Otherwise you could, for example, conclude erroneously, that $A \times B$ is always the same as $A + B$—since if you only looked at the specific case where $A = 2$ and $B = 2$, this would appear to be true. We hope that many of you would be interested, once you've become convinced of the truth of a rule like $(A \times B) \div C = A \times (B \div C)$ by examining some special cases, to try actually to prove the rule. In the last section of this chapter we show how such rules may be established.

$\frac{16}{64} = \frac{1}{4}$; $\frac{19}{95} = \frac{1}{5}$;

Therefore, $\frac{13}{33} = \frac{1}{3}$

Hm ?

Exercise 3.3

In Problems 1 through 5 carry out each of the computations in the order indicated by the parentheses and compare the two answers. You should not use (or need to use) a hand-calculator until you come to Problem 10.

1. (a) $(36 \div 12) \times 100$
    (b) $(100 \times 36) \div 12$

2. (a) $(15 \div 5) \times 10$
    (b) $(10 \times 15) \div 5$

3. (a) $(25 \div 5) \times 100$
    (b) $(100 \times 25) \div 5$

4. (a) $(56 \div 8) \times 10$
    (b) $(10 \times 56) \div 8$

5. (a) $(225 \div 15) \times 100$
    (b) $(100 \times 225) \div 15$

6. The results of Problems 1 through 5 illustrate a general rule. Complete the following two statements which formalize that rule:
    (a) $(A \div B) \times 100$ is the same as $(100 \times A) \div \underline{\ ?\ }$.
    (b) $(A \div B) \times 10$ is the same as $\underline{\qquad ? \qquad}$.

7. Use the technique described in the Remark at the end of Section 3.3 to determine which of the following general statements are *always* true.
    (a) $(A \div B) \times C$ is always the same as $(A \times C) \div B$.
    (b) $(A + B) - C$ is always the same as $A + (B - C)$.
         (Hint: In your examples choose values of C less than values of B.)
    (c) $(A - B) + C$ is always the same as $A - (B + C)$.

(d) $A \div B$ is always the same as $100A \div 100B$.

(e) $A \div B$ is always the same as $1000A \div 1000B$.

8. Perform the following computations.

(a) $4.8 \div 1.6$

(b) $.96 \div .16$

(c) $37.8 \div 4.2$

(d) $22.8 \div 1.9$

(e) $15.75 \div 1.05$

(f)

9. Make up a problem whose answer would 'fit' for part (f) in Problem 8.

10. First make a rough guess for the given division problem and then compute the answer using a hand-calculator if you wish. Record your answer with accuracy shown to two decimal places.

(a) $786 \div 42.3$ (Ans: Rough guesses will vary: for example $800 \div 40$ or 20. Computed answer 18.58)

(b) $46 \div 1.7$

(c) $53.7 \div 20.3$

(d) $856 \div 35.2$

(e) $.792 \div .25$

3.4 DIVISION OF FRACTIONS

The difficulty with division, which we stressed in the first two sections, is that it is not a well-defined operation of mathematics when applied to the whole numbers or decimals. There is no whole number (nor even decimal) that gives an exact answer to the problem $100 \div 3$. There is no decimal that gives an exact answer to the problem $50 \div 12.65$. From a *strictly mathematical* point of view, the great advantage of fractions is that it is always possible to give an exact, fractional answer to a division problem. This we shall soon see; but we are well aware that this advantage may have little or no appeal to practically-minded, intelligent adults, who seek a less 'theoretical' pay-off for the effort of learning how to divide fractions. So we will also show that the division of fractions can be useful—though we readily admit that the division of fractions arises much less frequently than the multiplication of fractions in real-life situations.

We already indicated one place where we need to divide fractions, when we discussed in Chapter 2 the relation of probability to odds and Bayes' Theorem giving us the posterior odds on a hypothesis in terms of the prior odds and the outcome of an experiment. How, then, should we divide a fraction by a fraction?

To answer this question, let us remember under what circumstances we *multiply* fractions. If we want to take, say, $\frac{2}{3}$ of $\frac{4}{5}$ (of something, for example, a length of cloth) then we multiply the

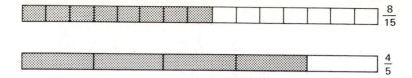

fractions to find what fraction of the cloth we are, in fact, taking. Moreover, we multiply by multiplying 'tops' and 'bottoms'. Thus

(4.1) $$\frac{2}{3} \text{ of } \frac{4}{5} = \frac{2}{3} \times \frac{4}{5} = \frac{2 \times 4}{3 \times 5} = \frac{8}{15},$$

so we are taking $\frac{8}{15}$ of the length of cloth. Suppose, instead, we ask ourselves the question, 'What fraction of $\frac{4}{5}$ is $\frac{8}{15}$?'† *We* know the answer is $\frac{2}{3}$, from equation (4.1), but how would we get the answer if we were just faced with the question?

First, we claim that, however we do the arithmetic, the process deserves to be called division; specifically, the division of $\frac{8}{15}$ by $\frac{4}{5}$. For the process of finding X in the equation, say

$$X \times 7 = 56,$$

is certainly that of dividing 56 by 7, so the process of finding X in the equation

(4.2) $$X \times \frac{4}{5} = \frac{8}{15}$$

should be called 'dividing $\frac{8}{15}$ by $\frac{4}{5}$'. So far so good; but we still have to say how we do, in fact, find X from the equation (4.2). We *could* proceed as follows. Suppose X is the fraction $\frac{N}{D}$. Then

$$X \times \frac{4}{5} = \frac{N}{D} \times \frac{4}{5} = \frac{N \times 4}{D \times 5}.$$

But we are told that $X \times \frac{4}{5} = \frac{8}{15}$, so‡

(4.3) $$\frac{4 \times N}{5 \times D} = \frac{8}{15}.$$

If the two sides of the equality (4.3) are indeed the *same fraction* then we have

(4.4) $$4 \times N = 8, \quad 5 \times D = 15.$$

Then, by ordinary division, we immediately get $N = 2$ and $D = 3$.

---

†It is conceivable that we might have $\frac{4}{5}$ of a bolt of cloth and wish to sell $\frac{8}{15}$ of a bolt. What fraction of what we have are we selling?

‡It makes no difference whether we write $\frac{N \times 4}{D \times 5}$ or $\frac{4 \times N}{5 \times D}$.

Thus

$$X = \frac{N}{D} = \frac{2}{3},$$

the right answer!

Have we solved the problem of dividing fractions by fractions? No, we must admit we have not. For our particular problem (4.2) was particularly easy—we were able to assume that, in (4.3), the two sides of the equality were the *same fraction*. Suppose, however, we were to try to apply this method to the problem

(4.5) $$X \times \frac{2}{3} = \frac{3}{4}.$$

Proceeding as before we would write $X = \frac{N}{D}$ and obtain

(4.6) $$\frac{N \times 2}{D \times 3} = \frac{3}{4}.$$

Now, however, it would be ridiculous to suppose that the two sides of the equality in (4.6) are the same fraction, because there is no whole number N such that $2 \times N = 3$ (nor a whole number D such that $3 \times D = 4$). What then can we do when faced with (4.6)?

We should now remember that it is not the *fraction,* strictly speaking, but the *rational number* represented by the fraction that comes into question in arithmetic—it is not often necessary to stress this precision but here it is essential. As numbers $\frac{2}{3}$ and $\frac{4}{6}$ are the same, though they are different fractions. Thus, in trying to solve Equation (4.5), we are free to change $\frac{2}{3}$ or $\frac{3}{4}$ into any other fractions representing the same rational numbers as $\frac{2}{3}$ or $\frac{3}{4}$, respectively. What fractions should we choose? Certainly, we have choice. Let us see if we can get away with only changing $\frac{3}{4}$. We can change it into $\frac{3 \times M}{4 \times M}$ for any whole number M. Our purpose should be that, in getting the new equation corresponding to (4.6) we now really can assume we have the same fractions on the two sides of the equality. We claim that we can take $M = 6$ (and we'll explain *why*, later). Then

$$\frac{3}{4} = \frac{3 \times 6}{4 \times 6} = \frac{18}{24}$$

and, since $X \times \frac{2}{3} = \frac{18}{24}$, we get, putting $X = \frac{N}{D}$

(4.7) $$\frac{2 \times N}{3 \times D} = \frac{18}{24}$$

so that we satisfy (4.7) if

(4.8) $$2 \times N = 18, \quad 3 \times D = 24,$$

that is, if

(4.9) $N = 9, \quad D = 8.$

We have thus solved our original problem (4.5) which is entirely equivalent to $X \times \frac{2}{3} = \frac{18}{24}$; we have found $X = \frac{9}{8}$, so that

(4.10) $\frac{3}{4} \div \frac{2}{3} = \frac{9}{8}.$

So far, so good—but how did we know, in our calculation above, that it would 'work' to take $M = 6$? Surely, we don't have to depend on inspired guessing! No, we don't! We had to choose a number M so that $3 \times M$ would be divisible by 2 and so that $4 \times M$ would be divisible by 3. This would certainly happen if M was divisible by 2 and by 3, that is, if M was divisible by 6—so we took the smallest such number M, namely, 6 itself.

We will describe one more example and then look for the general rule.

Example 1   Compute $\frac{4}{5} \div \frac{2}{7}$,

We are looking for $X = \frac{N}{D}$ so that $\frac{N}{D} \times \frac{2}{7} = \frac{4}{5}$. We must replace $\frac{4}{5}$ by an equivalent fraction $\frac{4 \times M}{5 \times M}$, so that we can achieve the equality

$$\frac{2 \times N}{7 \times D} = \frac{4 \times M}{5 \times M}$$

by actually making these the same fraction. If we follow the reasoning in studying (4.5), we will take $M = 2 \times 7 = 14$. (We could actually take $M = 7$. Why?) Then

$$\frac{4}{5} = \frac{4 \times 14}{5 \times 14} = \frac{56}{70},$$

so

$$\frac{2 \times N}{7 \times D} = \frac{56}{70},$$

and we satisfy this equation by taking

$2 \times N = 56, \quad 7 \times D = 70,$

or

$N = 28, \qquad D = 10.$

Thus $\frac{4}{5} \div \frac{2}{7} = \frac{28}{10}$. We can reduce our answer, $\frac{28}{10}$, to $\frac{14}{5}$ if we please.

Actually this example may be more instructive if we take a less ambitious approach and refuse to carry out any arithmetic computations, unless we are forced to do so. If we take that approach we are eventually faced with solving the equation

$$\frac{2 \times N}{7 \times D} = \frac{4 \times 14}{5 \times 14},$$

which will be satisfied if $2 \times N = 4 \times 14$ and $7 \times D = 5 \times 14$. You may easily verify by direct substitution that if $N = 4 \times 7$ and $D = 5 \times 2$ both of these equations are valid. Thus $\frac{4}{5} \div \frac{2}{7} = \frac{4 \times 7}{5 \times 2}$. (Can you observe an easy way to get the expression on the right from the numbers in the fractions on the left?)

We now proceed to develop the *general algorithm* for dividing fractions. Let the problem be $\frac{A}{B} \div \frac{Y}{Z}$. Thus we seek $X = \frac{N}{D}$ so that

$$X \times \frac{Y}{Z} = \frac{A}{B},$$

or

$$\frac{N}{D} \times \frac{Y}{Z} = \frac{A}{B}.$$

or†

$$\frac{Y \times N}{Z \times D} = \frac{A}{B}.$$

Remember that we are trying to find whole numbers, N, D making this true.

Our procedure is to replace the fraction $\frac{A}{B}$ by the equivalent fraction $\frac{A \times Y \times Z}{B \times Y \times Z}$, so that

(4.11) $$\frac{Y \times N}{Z \times D} = \frac{A \times Y \times Z}{B \times Y \times Z}.$$

If the two sides of (4.11) are to be exactly the same fraction, we must have

$$Y \times N = A \times Y \times Z, \quad Z \times D = B \times Y \times Z,$$

so that

$$N = A \times Z \qquad\qquad D = B \times Y, \text{ and}$$

---

†It makes no difference, when we are working generally, whether we write

$$\frac{Y \times N}{Z \times D} \text{ or } \frac{N \times Y}{D \times Z}.$$

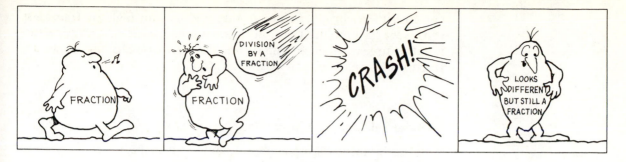

$$X = \frac{N}{D} = \frac{A \times Z}{B \times Y} \, .$$

Thus we have proved

(4.12)
$$\frac{A}{B} \div \frac{Y}{Z} = \frac{A \times Z}{B \times Y} \, .$$

If we look at the result in (4.12) we observe the following remarkable rule:

Division Algorithm for Fractions

To divide the fraction $\frac{A}{B}$ by the fraction $\frac{Y}{Z}$, turn the divisor $\frac{Y}{Z}$ upside down, getting $\frac{Z}{Y}$, and then multiply: thus

(4.13)
$$\frac{A}{B} \div \frac{Y}{Z} = \frac{A}{B} \times \frac{Z}{Y} = \frac{A \times Z}{B \times Y} \, .$$

Notice, however, that we have not produced the division algorithm 'out of the hat'. We have *demonstrated* that it gives the correct answer in situations in which we need to divide fractions. Thus, for example, the rule gives

$$\frac{3}{4} \div \frac{2}{3} = \frac{3}{4} \times \frac{3}{2} = \frac{9}{8} \, ,$$

just as in (4.10); and $\frac{9}{8}$ *is the right answer,* because

$$\frac{9}{8} \times \frac{2}{3} = \frac{9 \times 2}{8 \times 3} = \frac{18}{24} = \frac{3 \times 6}{4 \times 6} = \frac{3}{4} \, .$$

Finally, we remark that (4.13) shows that we can always divide a fraction by a fraction with absolute precision, getting again a fraction.† This is very satisfying mathematically, however cold the

†Of course we *can* divide a decimal by a decimal and get a fraction, since a decimal can always be converted into a fraction. But a fractional answer to a problem involving the division of decimals is seldom useful in practice. Likewise it is rarely best to give the answer to a problem involving fractions in decimal form though there may be cases where it is acceptable.

We can also divide mixed numbers (say, $3\frac{1}{2} \div 2\frac{1}{3}$ by converting them into fractions; we may, then, if we please, convert the resulting fractional answer back into the form of a mixed number.

comfort it may bring to some of our readers! But dividing fractions can, as we have pointed out, occasionally be useful in real-life, so it is better to learn—and to understand—the division algorithm for fractions.

Exercise 3.4

1. Use the division algorithm for fractions to obtain the answers to the following problems.

   (a) $\dfrac{1}{3} \div \dfrac{2}{5}$ (d) $\dfrac{4}{9} \div \dfrac{5}{11}$

   (b) $\dfrac{2}{5} \div \dfrac{3}{7}$ (e) $\dfrac{5}{11} \div \dfrac{6}{13}$

   (c) $\dfrac{3}{7} \div \dfrac{4}{9}$ (f) $\dfrac{6}{13} \div \dfrac{7}{15}$

2. (a) Look at the answers to the successive parts of Problem 1. Do you notice anything about their size (in relationship to the number 1)?

   (b) Write out the next three division problems that would fit in this sequence and compute the answers to see if your observation in part (a) still holds.

3. Carry out the following indicated divisions.

   (a) $4 \div \dfrac{3}{5}$ (Hint: $4 = \dfrac{4}{1}$)

   (b) $\dfrac{3}{5} \div 4$

   (c) $\dfrac{2}{3} \div \dfrac{7}{9}$

   (d) $\dfrac{7}{9} \div \dfrac{2}{3}$

   (e) $3\dfrac{1}{2} \div 2\dfrac{1}{3}$ (Hint: $3\dfrac{1}{2} = \dfrac{7}{2}$, $2\dfrac{1}{3} = \dfrac{7}{3}$)

   (f) $4\dfrac{1}{7} \div 7\dfrac{1}{4}$

   (g) $6\dfrac{1}{3} \div 3\dfrac{1}{6}$

4. What general rule is suggested by parts (e), (f) and (g) of Problem 3?

5. We already discussed in Section 1.3 how to divide a decimal number by a power of 10. In the following problems first convert the given numbers to their fractional representation, then use the division algorithm for fractions to compute the answer.

(a) $56.1 \div 10$ (Ans: $56.1 \div 10 = \frac{561}{10} \div \frac{10}{1}$

(b) $32.7 \div 100$ $= \frac{561}{10} \times \frac{1}{10}$

(c) $43 \div 10$ $= \frac{561}{100}$

(d) $213.2 \div 100$ $= 5.61.)$

6.   Perform the following indicated divisions.

(a) $\dfrac{3}{5} \div \dfrac{4}{4}$ (d) $\dfrac{6}{8} \div \dfrac{7}{10}$

(b) $\dfrac{4}{6} \div \dfrac{5}{6}$ (e) $\dfrac{7}{9} \div \dfrac{8}{12}$

(c) $\dfrac{5}{7} \div \dfrac{6}{8}$ (f) $\dfrac{8}{10} \div \dfrac{9}{14}$

7.   Observe that each of the answers to the various parts of Problem 6 is larger than its predecessor (you can convert to decimal approximations to verify that this is true, if necessary). Construct, and work, the next three problems that fit this sequence and see if this holds for these cases as well.

8.   The costume warehouse at Nonsense High School has $17\frac{1}{2}$ yards of cloth available to make costumes for the chorus in the school pageant. Each costume requires $\frac{3}{4}$ yard of cloth. How many costumes can be made?

9.   Refer to the facts in Problem 8. How much cloth was left over?

10.   Make up and solve some more problems like the last two.

*3.5 SOME ARITHMETIC IDENTITIES (LAWS OF ARITHMETIC)

At the end of Section 3 you should have become convinced, by the method of 'trial and error,' that the rule $(A \times B) \div C = A \times (B \div C)$ always holds. Such rules are often called 'identities.' Suppose though that we wished actually to *prove* a rule like this—how would we go about it?

First we should agree exactly what we mean by proving such an identity. For what numbers A, B, C are we claiming that the rule holds? Well, let us be ambitious and try to prove it for the most general sort of number we have so far introduced. These are the *rational numbers*;† that is, the numbers represented by fractions. For the natural numbers 0, 1, 2, 3, $\cdots$ are (special) rational numbers, since, for example $10 = \frac{10}{1}$; and decimals are also rational numbers since, for example, $3.14 = \frac{314}{100}$. We will now establish the rule $(A \times B) \div C = A \times (B \div C)$, for all rational numbers A, B, C (except that we must exclude $C = 0$ since we are not allowed to divide by 0), starting from the meaning of the multiplication of natural numbers.

†Of course, we speak here only of *positive* numbers (and zero).

Let us consider a rectangular box of length k inches, width ℓ inches, height m inches, where k, ℓ, m are natural numbers (if you like to have a definite example in mind, think of $k = 4$, $\ell = 5$, $m = 3$). What is the volume of this box?

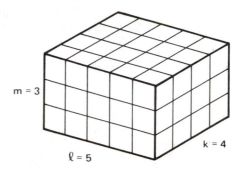

Well, its base area is plainly $(k \times \ell)$ square inches and its height is m inches, so its volume is $(k \times \ell) \times m$ cubic inches. On the other hand we may turn the box on its side and calculate its volume that way:

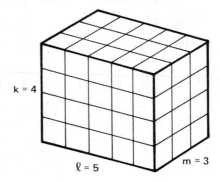

Its base area is now $(\ell \times m)$ square inches, so its volume is $k \times (\ell \times m)$ cubic inches. This shows that

(5.1)
$$(k \times \ell) \times m = k \times (\ell \times m)$$

for any natural numbers k, ℓ, m. We call (5.1) the *associative law of multiplication of natural numbers*. Notice that if we just look at the base area in the first picture, we immediately infer that

(5.2)
$$k \times \ell = \ell \times k$$

for any natural numbers k, ℓ. We call (5.2) the *commutative law of multiplication of natural numbers*. We should emphasize that we really have proved the laws (5.1) and (5.2) using only the meaning of multiplication of natural numbers,† since the volume of the box is, by definition, the number of cubes, 1 inch by 1 inch by 1 inch, that will fit into the box (and the area of the base is, by definition, the number of squares, 1 inch by 1 inch, that will fit into the base).

We now proceed to extend (5.1) and (5.2) to fractions. Recall from Chapter 2, Section 2, that we multiply fractions by the rule

$$\frac{a}{b} \times \frac{c}{d} = \frac{a \times c}{b \times d} \ .$$

Thus let X, Y, Z be three rational numbers, represented by the fractions $\frac{a}{b}, \frac{c}{d}, \frac{e}{f}$, respectively. Then

$$(X \times Y) \times Z = \left(\frac{a}{b} \times \frac{c}{d} \right) \times \frac{e}{f} = \left(\frac{a \times c}{b \times d} \right) \times \frac{e}{f}$$

$$= \frac{(a \times c) \times e}{(b \times d) \times f} \ ,$$

$$X \times (Y \times Z) = \frac{a}{b} \times \left(\frac{c}{d} \times \frac{e}{f} \right) = \frac{a}{b} \times \left(\frac{c \times e}{d \times f} \right)$$

$$= \frac{a \times (c \times e)}{b \times (d \times f)} \ .$$

But, according to (5.1), $(a \times c) \times e = a \times (c \times e)$ and $(b \times d) \times f = b \times (d \times f)$ (remember that a, b, c, d, e, f are all natural numbers!). Thus the fractions representing $(X \times Y) \times Z$ and $X \times (Y \times Z)$ have the same 'top' and the same 'bottom' and so are certainly the same. It follows that

(5.3) $$(X \times Y) \times Z = X \times (Y \times Z)$$

for all *rational* numbers X, Y, Z. This is the *associative law of multiplication of rational numbers.*

A similar, but simpler, argument shows that

(5.4) $$X \times Y = Y \times X$$

for all *rational* numbers X, Y. This is the *commutative law of multiplication of rational numbers.*

Now we proceed to discuss the identity with which we started this section. We saw in Section 4 that, to divide by the fraction $\frac{e}{f}$ is the same as multiplying by $\frac{f}{e}$. So suppose X, Y, Z are rational numbers represented by the fractions $\frac{a}{b}, \frac{c}{d}, \frac{e}{f}$ respectively. Then

$$(X \times Y) \div Z = \left(\frac{a}{b} \times \frac{c}{d} \right) \div \frac{e}{f} = \left(\frac{a}{b} \times \frac{c}{d} \right) \times \frac{f}{e}$$

$$= (X \times Y) \times W,$$

†You might claim that our argument doesn't cover the case when one of the numbers is 0— but this case is especially easy!

where W is the rational number represented by the fraction $\frac{f}{e}$ (it is, of course, natural, and customary, to write $\frac{1}{Z}$ or Z^{-1} for W). Similarly

$$X \times (Y \div Z) = \frac{a}{b} \times \left(\frac{c}{d} \div \frac{e}{f} \right) = \frac{a}{b} \times \left(\frac{c}{d} \times \frac{f}{e} \right)$$

$$= X \times (Y \times W).$$

But we know from (5.3) that ·

$$(X \times Y) \times W = X \times (Y \times W)$$

for all rational numbers X, Y, W. It follows immediately that

(5.5) $$(X \times Y) \div Z = X \times (Y \div Z)$$

for all rational numbers X, Y, Z, provided that Z is not zero.

There is an interesting and instructive point to be made about our method of obtaining the law (5.5). Suppose you had only been interested in proving (5.5) for *natural* numbers X, Y, Z; even then, to use the method we have described, you would still have required the associative law of multiplication of *rational* numbers, since W would not be a natural number unless $Z = 1$! Of course you could have found another way to prove (5.5) for natural numbers X, Y, Z, but the method we have described is *systematic* and *general*. We are not going to attempt any precise definition of 'systematic', but we hope you see what we mean!

Are there any other important laws of arithmetic? There certainly are. We will discuss only three others. First there is the law

(5.6) $$(X + Y) + Z = X + (Y + Z),$$

for all rational numbers X, Y, Z; this is called the *associative law of addition for rational numbers.* Second there is the law

(5.7) $$X + Y = Y + X,$$

for all rational numbers X, Y; this is called the *commutative law of addition for rational numbers.* Third, there is the law

(5.8) $$X \times (Y + Z) = (X \times Y) + (X \times Z),$$

3 × (4 + 5)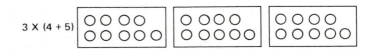

(3 × 4) + (3 × 5)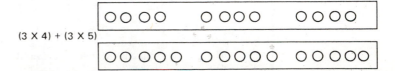

for all rational numbers X, Y, Z; this is called the *distributive law (of addition over multiplication) for rational numbers.* How would we prove these? We would proceed essentially as we did for laws (5.3) and (5.4). That is, we would first establish the validity of these laws for the *natural numbers*; this step, however, is easier than in the case of multiplication, and we will leave our readers to satisfy themselves that (5.6), (5.7), (5.8) are true if X, Y, Z are natural numbers. We then proceed to use the definitions of the addition and multiplication of fractions to extend our laws to rational numbers. We will be content to do this for laws (5.7) and (5.8)—the associative law of addition is not difficult, nor any different in principle, but it is more tedious to establish than (5.7) or (5.8).

Suppose then that we have established the identity (we certainly have!)

(5.9)
$$k + \ell = \ell + k$$

for natural numbers k, ℓ. Now if $X = \frac{a}{b}$ and $Y = \frac{c}{d}$, then, according to the formula at the end of Section 5 of Chapter 2,

$$X + Y = \frac{a}{b} + \frac{c}{d} = \frac{(a \times d) + (b \times c)}{b \times d} \; ,$$

$$Y + X = \frac{c}{d} + \frac{a}{b} = \frac{(c \times b) + (d \times a)}{d \times b} \; .$$

But, by (5.2) and (5.9)

$$(a \times d) + (b \times c) = (b \times c) + (a \times d) = (c \times b) + (d \times a)$$

and

$$b \times d = d \times b.$$

Thus $X + Y, Y + X$ are represented by the *same* fraction and so are certainly equal. This establishes (5.7).

To deal with (5.8) we again suppose we have established the distributive law for the natural numbers,

(5.10)
$$k \times (\ell + m) = (k \times \ell) + (k \times m).$$

This is achieved by studying the following picture where, by way of example,

$$k = 3, \quad \ell = 2 \quad \text{and} \quad m = 5.$$

Now if

$$X = \frac{a}{b}, \quad Y = \frac{c}{d}, \quad Z = \frac{e}{f}$$

then

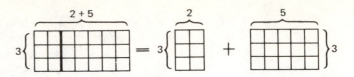

$$X \times (Y + Z) = \frac{a}{b} \times \left(\frac{c}{d} + \frac{e}{f} \right) = \frac{a}{b} \times \frac{(c \times f) + (d \times e)}{d \times f}$$

$$= \frac{a \times \{(c \times f) + (d \times e)\}}{b \times (d \times f)} \quad .$$

We now invoke the distributive law for the natural numbers (5.10) to infer that

$$X \times (Y + Z) = \frac{(a \times (c \times f)) + (a \times (d \times e))}{b \times (d \times f)} \quad .$$

On the other hand,

$$(X \times Y) + (X \times Z) = \left(\frac{a}{b} \times \frac{c}{d} \right) + \left(\frac{a}{b} \times \frac{e}{f} \right) = \frac{a \times c}{b \times d} + \frac{a \times e}{b \times f}.$$

Now $b \times (d \times f) = (b \times d) \times f$, and $b \times (d \times f) = b \times (f \times d) = (b \times f) \times d$; here we use (5.1) and (5.2). Thus $b \times (d \times f)$ is a common multiple of $(b \times d)$ and $(b \times f)$ (you probably realized this already!) and

$$\frac{a \times c}{b \times d} = \frac{(a \times c) \times f}{(b \times d) \times f}, \frac{a \times e}{b \times f} = \frac{(a \times e) \times d}{(b \times f) \times d} = \frac{(a \times d) \times e}{(b \times d) \times f},$$

by further applications of (5.1) and (5.2). Thus

$$(X \times Y) + (X \times Z) = \frac{((a \times c) \times f) + ((a \times d) \times e)}{(b \times d) \times f}$$

and it is plain, applying (5.1) yet again, that

$$X \times (Y + Z) = (X \times Y) + (X \times Z),$$

as claimed.

You may well wonder why it was more complicated to establish (5.7) and (5.8) than (5.3); the reason is that the *multiplication* of fractions involves only the *multiplication* of natural numbers, whereas the *addition* of fractions involves both the *addition* and *multiplication* of natural numbers. The addition of fractions really is more difficult than the multiplication of fractions; mercifully, it is also less important:

We should surely also be able to prove the analogous result to (5.5), using addition and subtraction in place of multiplication and division; that is we certainly believe the law

(5.11) $$(X + Y) - Z = X + (Y - Z)$$

for all rational numbers X, Y, Z. However, we are not (yet) able to produce an argument similar to the one whereby we derived (5.5) from (5.3) *because we don't yet have negative numbers* (Chapter 5). Do you see that negative numbers stand in the same relation to natural numbers under *addition* as rational numbers stand to natural numbers under multiplication? Think about this if it is not already clear. We introduce rational numbers so that we can always 'undo multiplication', that is divide; and we will introduce negative numbers so that we can always 'undo addition', that is, subtract. Moreover, once we have rational numbers, division by N becomes multiplication by the inverse of N and, once we have negative numbers, subtraction by N becomes addition by the negative of N. Thus, as an example of the 'inverse' or 'undoing' of multiplication, notice that if we divide a number (say 24) by another number, say 3, the result is exactly the same as multiplying that number by the inverse of 3, which is $\frac{1}{3}$, or 3^{-1}. This may be written as

$$24 \div 3 = 8, \quad \text{or } 24 \times \frac{1}{3} = 8, \quad \text{or } 24 \times 3^{-1} = 8.$$

Similarly, when we introduce negative numbers in Chapter 5 you will see that

$$8 - 5 = 3, \quad \text{or} \quad 8 + (-5) = 3, \quad \text{or} \quad 8 + (-5) = 3;$$

where the '$-$' sign is used in the first instance to mean subtraction and the second two instances to signify that the number to the right of that symbol is to be regarded as the additive inverse of that number (the '$-$' sign appearing in the exponent of the multiplication example has exactly the same meaning).

But, since we have not yet developed negative numbers we must seek a different proof of (5.11) if we wish to prove it now. We must, moreover, restrict X, Y, Z, requiring that $Y \geq Z$, so that all the numbers entering into our discussion are positive. So let us suppose that X, Y, Z are (positive) rational numbers with $Y \geq Z$. Let $Y = U + Z$ so that $U = Y - Z$. Then

$$(X + Y) - Z = (X + (U + Z)) - Z = ((X + U) + Z) - Z, \text{ by (5.6)}$$
$$= X + U, \text{ by definition of subtraction as undoing addition}$$
$$= X + (Y - Z),$$

and (5.11) is proved if $Y \geq Z$.

CHAPTER

4

Estimation and Approximation

This topic, often neglected at this level, is of vital importance in the everyday use of mathematics and in the intelligent use of calculating machines. We relate estimation to the elementary study of statistics.

4.1 APPROXIMATION AS A PRACTICAL TOOL

We already saw in Chapter 1 that the use of decimals to record measurements really implies that the measurements are only made to a given degree of accuracy. Thus, if we are told that a certain distance AB is 23.62 meters, this means that an accuracy to within the nearest centimeter is claimed for the measurement of AB. If we are asked for a quarter of the distance AB, it would be absurd to give the answer 5.905 meters, since we certainly cannot guarantee the figure 5 representing a millimeter reading. Thus an exact arithmetical calculation is here strictly irrelevant to the problem at hand; what we need is an *approximate* calculation—we would probably be content to say that the quarter-distance is between 5.90 meters and 5.91 meters.

We will discuss examples of this kind in greater detail in Section 3, after a more careful treatment of accuracy of measurement. In this section we will give you many further examples in which a problem is solved with the help of an approximate calculation; in these examples, as in the example above, an accurate calculation would be irrelevant to the solution of the problem or, at best, a step towards the solution and perhaps not the most efficient step.

Before taking up these examples, let us bring out into the open one of the most popular misconceptions about mathematics among those who don't really understand it. According to this mistaken belief, mathematics, being 'certain' and 'precise', cannot deal in approximate answers and cannot countenance a number of different and equally valid solutions to a problem.†

The mathematical theory of probability, of which we have already said quite a lot in Chapter 2, is a full-fledged component of mathematics, dealing with uncertainties. There is also a mature area of mathematics called *approximation theory* which is very active and, of course, full of interesting applications. We will not take the reader far into this theory in this volume or its successors, but we will present, in this section, some examples of approximate calculation which do not proceed via a preliminary exact and accurate calculation. Moreover, we need hardly emphasize that, when making approximations, choice is exercised and no single choice is to be picked out as the single, unique right answer. You should already be familiar with the fact that (contrary to the prevailing misconception) there are usually many different ways of solving a mathematical problem; now we want to emphasize that approximate calculations may have different valid answers, quite apart from the method of solution. Thus if we want the approximate area of a field 12.3 meters by 13.7 meters, then we must admit 170, 168, 169, 168.5 square meters as valid answers. *Which* answer we choose will depend on the purpose, or context, of the calculation; thus, it is especially important to take account of the actual question we want to answer in deciding our strategy with regard to an approximate calculation or estimation. It is important to understand that an approximate calculation can lead to a perfectly definite answer to a question.

Let us give some examples:

(1)   **Can I afford it?** This type of question occurs frequently (indeed, as the state of the economy declines, with increasing frequency) in real life. I have $10.00 with me when I go to the supermarket and I want to buy a large package of detergent ($6.37), a pound of cod ($1.69) and a dozen eggs ($0.85). Do I have enough money? It is not necessary to do a precise calculation to answer this question. Another problem of this kind would be: I receive a raise of $350 per month. I would like to send my daughter to a private school, but the fees are $3000 a year. Can I afford it? The answer here is yes—assuming I was already coping with my budgetary problems before my raise, and that I have not overlooked any hidden expenses involved in my daughter going to the new school.

---

† A further part of this erroneous view of mathematics is that *every* mathematical problem has a solution. This is wrong, practically and philosophically, but it is not this particular error with which we are here concerned.

(2) **Do I have enough gas?** There was a time when we could travel by car without worrying about getting gas when we needed it. At whatever hour of the day or night, whether on the weekend or on a weekday, there would always be gas stations open. Today this is no longer the case, so that we have to be far more prudent and careful in estimating whether we can complete our journey given the amount of fuel in the tank (not to speak of that other estimate we also must make—can we afford the trip at all!?). Many factors enter into such an approximate calculation, of which obvious ones are our expected speed, the rate of consumption of gas at that speed, the distance to our destination (or to some reliable source of gas), and the amount of fuel in the tank. There may also be considerations of time as well, since we may need to take into account the times of day at which we may expect to find gas stations open. At certain times of crisis there is the additional hazard created by the possible unavailability of gas. We may need to know how many gallons our tank holds, but this may not be necessary information; for we may have recorded the fuel consumption using the tank capacity of our car, rather than the gallon, as our unit of volume. With all the information available—and the information will consist of estimates and approximations—we can decide upon a rational strategy: should we fill up now, even though the tank is nearly three-quarters full? The point we are making is that the answer may be an absolutely definite 'yes' (or 'no'), even though it is based on an approximate calculation employing estimated data. Exercises in such decision-making will be found at the end of this section.

(3) **Which is the better buy?** We are very often faced with questions of this kind, in our day-to-day shopping at the supermarket, as in the more critical decisions when purchasing a car or planning a vacation. If we want to buy detergent, we have to consider different brands and containers of different size. Thus we have to do some arithmetic in deciding just what quantity of what brand to purchase—and to do this arithmetic we must first collect data, i.e., we must find out the prices and the volume, or weight, of the contents of the various containers.† However, the calculation that we need to do in deciding the better buy is usually an approximate calculation rather than an exact one—we are comparing two unit prices and

---

†It is a popular fallacy that the larger container always gives us the better buy. We have come across several cases where this is not so. Moreover, it often happens that the larger container produces more waste, so that we lose a lot of value in buying it.

we want to know which is the smaller—we will usually not need to know exactly how much smaller. For example, if a packet of detergent of one brand weighs 5 lbs. and costs $2.60, while a packet of detergent of another brand weighs 3 lbs. and costs $1.80, then the price per pound of the smaller packet is 60¢ and that of the larger packet is less than 60¢, so that the larger packet is, apparently, the better buy. We are bound here, in all honesty, to add the qualification 'apparently' because, of course, we have by no means taken into consideration all the factors which we would like to incorporate into our decision-making—for example, which detergent does the better job? In choosing between different canned foods these considerations of quality become very important (for example, the amount of water in different brands of wet dog food may vary very substantially)—and finally, we want our food to taste good, too! But at least we can form a reasonable quantitative judgment using approximate calculations and—in the case of vacation planning, for instance—reasonable estimates.

(4) **Who has the better average?** We judge the merits of athletes (and others) by representing their performance as a certain average—for example, a baseball player's average is the ratio of the number of hits to number of times at bat. Though we argued in Chapter 2 that we use averages too much and sometimes in very misleading ways, nevertheless, such averages are part of our normal 'outlook on life' and often have to be compared. To say that player A's

average is better than player B's is to assert that the fraction corresponding to A's average is bigger than that corrsponding to B's. Thus we need to be able to take two fractions and say which is the bigger. This does *not* require that we be able—and willing—to subtract fractions, unless we want to say by how much A's average is superior to B's (and, in fact, the *differences* between their averages is not the best way to answer the 'how much' question). We can do an approximate calculation to convert the fractions into approximating decimals, as a means of deciding which fraction is bigger; or we can use the rule that the fraction $\frac{A}{B}$ is bigger than the fraction $\frac{C}{D}$ if and only if $A \times D$ is bigger than $B \times C$, and then approximate $A \times D$ and $B \times C$. In any case, the question will practically always be answered without having to do a precise calculation.

(5) **Which car gets the better mileage?** This type of question, in some ways related to that discussed under (2) above, may also be regarded as a special case of (3) and a variant on (4). However, it presents very special features, because we must take into consideration the circumstances under which the cars we wish to compare will be driven. One car may give the better mileage on the long trip, the other in city driving; one car may give better mileage in hilly country, the other on the flat. Moreover, we may well wish to consider not only mileage but also comfort, reliability, availability of spare parts, and the type of gas the engine uses. In the exercises we will exemplify this type of decision-making and the approximate calculations that may be called for.

(6) **Is the task possible?** Another type of problem calling for approximate calculation and, very often, estimation has to do with the feasibility of a task or project. We may wish to make a dress out of a certain amount of material—or, perhaps, 20 identical dresses from a certain amount of material. We may wish to put our dirty clothes through the washer and dryer before our favorite television program begins. We may wish to charge a number of our purchases to our credit card without exceeding the credit limit. All these questions, and others like them, involve us in approximations.

(7) **What is the next number below, the next number above?** In Section 2 of Chapter 3 we described a number of different real-life situations giving rise to the division of whole numbers. We showed there how, faced, for example, with the necessity of dividing 113 by 6, the correct answer to

the real-life problem might be obtained by taking the largest whole number less than $\frac{113}{6}$, namely 18; and, in other situations, the correct answer might be obtained by taking the smallest whole number greater than $\frac{113}{6}$, namely 19. When the problem requires one or the other of these approximations to the exact quotient of 113 by 6, it is plain that we should use approximation techniques, rather than exact algorithms, to get the answer. Notice that the answer to the question 'How many round trips, each costing $6.00 can I make if I have saved $113?' is *exactly* 18, but it is obtained by means of an *approximate* calculation.

We close this section by drawing your attention to a very different but extremely important use of approximation and estimation. Indeed, this use will doubtless be of increasing importance as we advance through the decade. We refer to the need to carry out checks on calculations done on the hand-calculator. Thank goodness we no longer need to do dreary calculations by hand, using uninteresting algorithms. If we need to do an accurate calculation (as we do, for example, in working out our federal income tax), then we may resort to a machine to take away the drudgery, but we would be well advised to check our accurate calculation by an approximate one done by hand. For, faced with the problem 81.2×7.4, if our calculator produces the answer 6008.8, we know it must be wrong, since 80×7 is 560. In this case, we must have made a mistake in punching the buttons. This is a common kind of error and approximate calculations will detect it. Notice that we are speaking here of a check on the *reasonableness* of an answer—of course, an approximate calculation cannot guarantee us the absolute correctness of the outcome of a *precise* calculation. This use of approximation methods also illustrates the point we made at the beginning of this section that there may be several different approximations all equally well serving the purpose of establishing the reasonableness of our answer. The validity of different approximations is also exemplified in several of the situations described in this section—but not, notice, in the situation of (7).

We will not discuss the actual arithmetical techniques involved in approximate calculations until Section 3; our next section discusses how to approximate numbers and introduces scientific notation for presenting numbers, determined to a required degree of accurary, in a particularly convenient way.

Exercise 4.1

1. Without doing an exact calculation determine which of the following shopping lists you can afford if you have only $20.00 to spend.

    (a) 2½ dozen eggs for $2.05
         1 gallon of corn oil for $6.95
         1 beef pot roast for $7.50
         3 boxes of tissues at 97¢ per box
         1 can of apple juice for $1.65
         1 package of candy at $1.25

    (b) 3 packages of chocolate chips at $2.04 per package
         1 six-pack of soda pop at $1.85
         4 bottles of shampoo at $2.80 per bottle

    (c) 1 bottle of instant coffee at $5.75
         6 dozen eggs at 85¢ per dozen
         2 pounds of bacon at $1.70 per pound

    (d) 10 cans of dog food at 87¢ per can
         5 cans of pineapple at 60¢ per can
         4 boxes of sandwich bags at $1.20 per box
         2 bottles of wine at $3.49 per bottle

2. (a) Suppose you own a stock that begins paying you a dividend of $30.00 quarterly (four times per year) and you decide to spend this money on subscriptions to magazines. If you want to subscribe to exactly three magazines and pay for the subscriptions with your newly acquired dividend income, which magazines might you choose?

Magazine	Price
See	$3.50 per month
Talk	$1.25 per week
Redwood Quarterly	$5.00 per quarter
Washington Pillar	$3.25 bi-weekly
Musical Reviews	$2.50 per month

    (b) Under these conditions which, if any, would you not be able to subscribe to at all?

3. Suppose service stations are open only between 10:00 AM and 2:00 PM and you are driving a car that averages 15 miles per gallon and has a 22-gallon tank. Also assume that you average 50 miles per hour (this allows time for brief stops along the way). Is it possible, under these circumstances, to make

    (a) a 400-mile journey in one day?

    (b) a 500-mile journey in one day?

    (c) a 1000-mile journey in one day?

4. Suppose you plan a 900-mile trip, say between San José and Seattle, in a car that gets 12 miles to the gallon, and gasoline for your car costs 129.9¢ per gallon. If you must stop one night at a motel (about $30.00) and buy three meals (for a total of about $20.00), would it be cheaper to take the airplane, if the airfare between those two cities is $115.00?

5. Here are some prices taken from the shelves of our local store. In each case determine which brand, or size, is the better buy.

(a)   Paper tissues

Brand	No. sheets	Ply	Price
W	280	2	$1.05
X	134	3	77¢
Y	200	2	59¢
Z	200	2	73¢

(b)   Aluminum foil

Size	Price
25 sq. ft†	53¢
200 sq. ft	$3.37

(c)   Spray laundry soap

Size	Price
16 oz aerosol spray can	$1.59
22 oz non-aerosol hand pump bottle	$1.69

(d)   Tub and shower cleaner

Size	Price
6 oz	$1.35
12 oz	$1.69
18 oz	$2.83

(e)   Potato chips

12 packages each weighing $\frac{1}{2}$ oz for $1.60
1 package weighing 8 oz for $1.13.

(f)   Medium AA eggs

Quantity	Price
1 dozen	83¢
$2\frac{1}{2}$ dozen	$2.05

6.   Liquid detergent is offered at a certain supermarket at the following prices.

Brand	Size (in oz)	Price
A	12	71¢
A	22	$1.15
A	32	$1.75
B	12	71¢
B	22	$1.25
B	32	$1.65
B	48	$2.55

(a)   Which size of the Brand A detergent is the better buy?

(b)   Which size of the Brand B detergent is the better buy?

(c)   Would you choose Brand A or Brand B on this evidence?

(d)   What other factors might you take into consideration in making your choice?

7.   Write a problem for each of the situations (1) through (7) described in this section. Then describe how you would solve the problem and give the solution.

†There was a sign next to this item that asked, "Why pay more, COMPARE, do yourself a favor!" There is no doubt that this advice was good, we compared and bought the *other* size.

4.2 ROUND-OFF; SCIENTIFIC NOTATION

It may, of course, very well happen that we are given a decimal number to a greater degree of accuracy than we require and we wish to approximate to that number. There are two basic reasons why this situation arises and it is important to distinguish between them. On the one hand, it may be quite unnecessary to record a number, or the result of a calculation, to the degree of accuracy available. We have given many examples of this situation in the previous section, where the decision to be taken, or the comparison required, does not necessitate great accuracy. If a magazine costs $2.25 a week and there is an 11% sales tax and we want to know if the annual cost exceeds $120, it is unnecessary to calculate that 11% of $2.25 is exactly 0.2475 dollars—we would 'round-off', probably, to 25 cents, obtaining the figure of $2.50 for the weekly outlay and deduce that our $120 is not quite enough.

The answer is roughly 4.2857142

On the other hand, there are situations in which the accuracy we have achieved in a calculation is really quite spurious. This happens very frequently today when we use hand-calculators so extensively. For a hand calculator always gives us answers to at least 7 places of decimals, and our measurements will very often not warrant such accuracy. Let us give some examples, too, that are independent of the hand-calculator. If we want to mark out a square plot of area roughly 2 km^2, then the length of a side can be calculated as 1.414 km. But it is absurd to aim for accuracy to the nearest meter when we're dealing with a length like 1.4 km, so we would be—and should be—perfectly content with an accuracy of 1.4 km. If we want to know the area of a rectangular field 1.26 km by 0.81 km it would be misleading to announce that the area is 1.0206 km^2, since we only know the length of the sides to the nearest dekameter (= $\frac{1}{100}$ km). Thus a sensible report of the area would be 1.02 km^2.

Thus there are situations in which we round off in view of the accuracy we *need* and situations in which we round off in view of the accuracy we can genuinely *get*—and, of course, situations in which both kinds of consideration apply. Is there then a technique of 'rounding off'? The answer is that there certainly is, and it is easy, but it depends on a difficult process that requires experience and judgment, and that cannot be rendered automatic except in very elementary sitautions, namely that of deciding the appropriate level of accuracy. We will give examples in the exercises of opportunities for such judgment; here, in the text, we will assume that we have already decided what level of accuracy we need (or are entitled to on the basis of the measurements we have made), and describe the rounding-off process.

The level of accuracy desired may be expressed in terms of a unit measure, thus, 'to the nearest cent', 'to the nearest millimeter', 'to the nearest gram', etc. However, if our problem has been translated into arithmetical terms, then the level of accuracy will itself be translated into purely numerical terms. Thus if the unit cost of some item is $12.44 (to the nearest cent, or as a precise amount) and we record the amount in *dollars,* we will obtain the number 12.44 *to the nearest hundredth,* or *to two places of decimals.* Again, if the weight of some object is 3.28 kg approximately, then we may record the weight in kilograms as 3.28 to the nearest hundredth or to two places of decimals. If we had a particularly sensitive weighing device we might feel confident in announcing the weight as 3.282 kg and the corresponding number would be 3.282 to three places of decimals. Of course we might, in this last case, choose to report the weight in *grams* instead of kilograms, in which case the number recorded would be 3282 to the nearest whole number. Sometimes it is more natural to describe the accuracy in terms of the number of *significant figures,* rather than the number of places of decimals— this allows for changes of basic unit, as in the example above. However, we must then be careful to specify, in giving a whole number ending with zeros, such as 24600, to *state* whether the zeros are significant (i.e., reliable) figures. So 24600 could contain three, four or five significant figures.

Thus, in a rounding-off problem, the data consist of a number together with a *level of accuracy* stated in the form 'to the nearest hundred', 'to the nearest whole number', 'to the nearest thousandth', etc. How then do we execute the rounding-off process? Well, given a number such as 23.5794,

if we round off to the nearest ten, we get 20;
if we round off to the nearest whole number, we get 24;
if we round off to the nearest tenth, we get 23.6;
if we round off to the nearest hundredth, we get 23.58;
if we round off to the nearest thousandth, we get 23.579.

The technique should be obvious—we find the (whole number) multiple of the stated level of accuracy closest to the number given, and this we do by cutting off digits to the right. There are certain special points to be noted however.

(a)   If the piece cut off starts with any of the digits 5, 6, 7, 8 or 9, we must *increase* the approximation by 1 in the last (right-hand) surviving place† (as we did to get 24, 23.6, and 23.58);

(b)   If we require accuracy to the nearest ten, hundred, thousand, and so on, then zeros will inevitably appear in our round-off approximation (as in the case of 20 above);

†There is a small exception to this rule: if what we cut off is *exactly* 5, we may increase or not, according to our choice.

(c) If we round off 17.01, say, to the nearest tenth, we must announce the result as 17.0, *not* 17. For if we suppress the zero after the decimal point, then our answer would be interpreted as approximating to the nearest whole number. Thus, although 17 and 17.0 are equal as *numbers,* they are not equivalent as approximations (just as 17 m and 17.0 m are not equivalent as measures of length).

(d) (This point is important and even though it may have occurred to you it should be stated formally.) If the number N is to be rounded off, it cannot be rounded off to an accuracy greater than that to which the number N is itself given. Round off 26.3 to the nearest tenth—answer 26.3. Round off 26.3 to the nearest hundredth—answer IMPOSSIBLE!

So far we have discussed rounding off to the *nearest* tenth (hundredth, ten, etc.). However, there are situations when we need to round *up* or round *down.* For example, if a supermarket is selling cans of sardines at $1.61 for two cans, then you may be sure that you would have to pay 81¢, not 80¢, for one can. Thus in working out unit prices where the price is quoted in bulk, we always round *up.* You have probably noticed that in the market place this rounding up happens even when the number determined by the appropriate arithmetic is really closer to the lower number. Thus if apples are priced at 4 for 61¢, then even though 61 divided by 4 is 15.25 (certainly closer to 15 than to 16) you would be charged 16¢ for one apple.

If, on the other hand, we are told to buy 6 items of a certain kind and not to spend more than $10, then, to determine the maximum unit price we can afford, we carry out the required division obtaining an answer (if we use the hand-calculator) of 1.6666667 and we *must* round down to $1.66. The context of an approximation will determine whether we will need to round up, round down or round off. If we do need to round *down,* it should be clear that we simply carry out the process above, except that our special point (a) should be ignored. If we need to round *up* we modify the process by always increasing the approximation by 1 whenever there are non-zero digits in the piece cut off. Thus, rounding down 27.681 to the nearest tenth, we get 27.6; rounding up 161.24 to the nearest whole number, we get 162. Remember then that we speak of three different kinds of rounding processes:

rounding *off,*   rounding *up,*   and rounding *down.*

We will discuss in Section 4.3 the special arithmetic that applies to computations which involve various combinations of rounded numbers. However, first we need to take up a very important question: is there a really convenient way of presenting approximate numbers, so as to facilitate their understanding and their arithmetic?

Scientists and engineers have indeed arrived at an agreed opinion on this—the result is what we call *scientific notation.* If, for example, we wish to present the approximation 32.58 in scientific notation, what we do is to rewrite the number with the decimal point following the first digit on the left, that is, we write 3.258; and then we 'adjust' by multiplying by a suitable power of 10, in this case, 10^1. Thus, in scientific notation,

$$32.58 = 3.258 \times 10^1 \ (= 3.258 \times 10).$$

Similarly,

$$3987 = 3.987 \times 10^3,$$

$$46910000 = 4.691 \times 10^7.$$

(Notice that the power of 10 turns out to be just the number of places we have moved the decimal point to the left.)

So far, so good—but what would we do if we want to write a number smaller than 1 in scientific notation? Of course, we could simply use a similar device but then *divide* by an appropriate power of 10. Thus, for example, we have the equalities

$$0.0284 = 2.84 \div 10^2.$$

$$0.00050 = 5.0 \div 10^4.$$

However, this is not quite what is understood by scientific notation. There are very good reasons, based on the so-called *law of exponents*

$$X^a \times X^b = X^{a+b},$$

and explained in detail in Section 5.2, for regarding 10^0 as a sensible symbol for the number 1, 10^{-1} as a sensible symbol for $\frac{1}{10}$, 10^{-2} as a sensible symbol for $\frac{1}{10^2}$, and, in general 10^{-a} as a sensible symbol for the fraction $\frac{1}{10^a}$, where a is any whole number.† Thus, since dividing by 10^2 is the same as multiplying by $\frac{1}{10^2}$, we may rewrite our last two equalities as

$$0.0284 = 2.84 \times 10^{-2},$$

$$0.00050 = 5.0 \times 10^{-4},$$

and now the expressions on the right are expressed in scientific notation. This is *all* we will do with negative numbers in this chapter. It is in the next chapter that we discuss their arithmetic.

What then, more explicitly, are the advantages of scientific notation? We claim that an expression like 4.691×10^7 is easier to

---

†We, in fact, introduced Z^{-1} as a notation for $\frac{1}{Z}$ in Chapter 3.

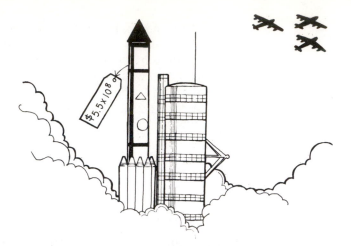

grasp and to manipulate than 46910000. Having the exponent 7 presented as part of the notation enables us to see immediately the *order of magnitude* of the number. We have met people in this country who cannot remember whether the defense budget is 126 million dollars or 126 billion dollars.† Thus they have remembered the relatively insignificant figure 126 but failed to remember the much more significant *order of magnitude*. We contend that had they been familiar with scientific notation it is very unlikely that they would have confused 1.26×10^8 with 1.26×10^{11}. Moreover, scientific notation makes it considerably easier to make comparisons between numbers. Thus if the defense budget is regarded as sacred— indeed, it is so sacred it is constantly increasing—it is easy to see why we have been so unsuccessful in our attempts to reduce government spending and thereby slow down inflation. For, whenever we make inroads in our social services, for example, we are dealing with expenditures measured in millions (10^6), while the defense budget is measured in billions ($10^9 = 10^3 \times 10^6 = 1000 \times 10^6$).

Thus in scientific notation we exhibit the *order of magnitude* (10^{11} in our defense budget example) and the *significant figures*†† (1.26 in that same example). Multiplication and division of such approximations or estimates are thereby also facilitated, since we multiply or divide powers of 10 by using the law of exponents, and we may easily estimate the result of multiplying or dividing numbers between 1 and 9, as a check on a computation on our hand calculator. We will give examples of these concepts in the next section.

We close this section by remarking that we are certainly *not* suggesting that every time you make an approximation or an estimate

---

†Remember we are writing this in 1979–80. The defense budget is now much higher than $126,000,000,000.

††We wrote $0.00050 = 5.0 \times 10^{-4}$ rather than 5×10^{-4}, in our example on page 155, because the zero after the decimal point is, here, a *significant* figure.

you should record it in scientific notation. Scientific notation should be used for quantities you may need to remember, especially if they are very large or very small; or for quantities on the basis of which you wish to make further estimates. If it is just a matter of some transitory interest, some approximate calculation which, once done, does not need to be recorded in your memory or used for some further arithmetical purpose, then by all means you should feel free to keep it in the form in which it naturally arose. We should always use the notation best adapted to our purpose; if scientific notation simplifies the execution of that purpose, it should be used. It is important, however, to get used to scientific notation so that you have a good feel for orders of magnitude.

Exercise 4.2

In Problems 1 through 7 round off the numbers to the nearest ten, nearest whole number, nearest tenth and nearest hundredth. If the rounding makes no sense, because the number hasn't been given to sufficient accuracy, simply write 'impossible'.

1.   42.1   (Ans: 40, 42, 42.1, impossible to round to the nearest hundredth.)

2.   188.764

3.   198.764

4.   8.427

5.   .015

6.   98000 (with five significant figures)

7.   98000 (with two significant figures)

In Problems 8 through 12 round the number *down* to the nearest ten, *down* to the nearest whole number, and *down* to the nearest tenth.

8. 17.34

9. 358.27

10. 182.11

11. 180.01

12. 38.41

In Problems 13 through 17 round the number *up* to the nearest ten, *up* to the nearest whole number, and *up* to the nearest tenth.

13. 17.34

14. 358.27

15. 182.11

16. 180.01

17. 38.41

18. Can you ever get the same answer by rounding a number

(a) up and off?

(b) down and off?

(c) up and down?

19. Suppose a number, rounded down to the nearest tenth, is 180.0. Give an argument why that number rounded *up* to the nearest unit should be 181.

20. Describe a situation in which it would be sensible to round 147.2 up to 148; and a (different!) situation in which it would be silly.

21. Each number in the column on the left is equivalent to one or more of the numbers in the column on the right. For each number in the column on the left, write that number followed by all of the equivalent numbers that appear in the right hand column. If any of the numbers are in scientific notation, circle them.

38	38×10^{-1}
3.8	3.8×10^1
380	3.8×10^{-2}
.0038	3.8×10^2
3800	38×10^{-4}
.038	$.038 \times 10^5$
	3.8×10^{-3}
	3.8×10^3

4.3 APPROXIMATION AS AN ARITHMETICAL OPERATION

Suppose the odometer on our bicycle is calibrated in tenths of a kilometer. We note the reading when we leave the house—it is 263.6. When we arrive home again the odometer reads 287.2. How far have we cycled? The obvious answer is $287.2 - 263.6$, or 23.6 km, but can we be sure of the accuracy of the last digit, 6? Let us study the situation a little more closely. The reading 263.6 actually means that the bicycle has been ridden between 263.6 and 263.7 kilometers

when we started out; and the reading 287.2 means that the bicycle had been ridden between 287.2 and 287.3 kilometers when we returned. Thus the *shortest* distance we might have traveled is 287.2 − 263.7 = 23.5 km, and the longest distance we might have traveled is 287.3 − 263.6 = 23.7 km. Thus we cannot be sure we have ridden 23.6 km (to the nearest tenth of a kilometer, or hectometer); we might have ridden anything from 23.5 to 23.7 kilometers. It is easy to see (you will have a chance to go through the argument in the Exercises) that 23.6 remains the *most likely* estimate (to one place of decimals) of the distance we have traveled, being much more likely than 23.5 or 23.7, which are equally probable. How then should we answer the question 'How far did you cycle today?' The best answer would be '23.6 km, but it could be 23.5 or 23.7, to the nearest tenth of a kilometer'. This kind of situation, and the calculations involved, are typical of a subtraction problem based on measurements which have been rounded down (as in the case of an odometer reading) or rounded up, where we wish to present the answer *rounded off.* Let us now give an example of an addition problem involving rounding up, where it is not only natural but actually correct in the context also to give the answer *rounded up.*

Suppose there is a regulation in a school district that, on any outing, there must be an adult to accompany every ten children. Thus, for example, if 83 children go on an outing they must be accompanied by 9 adults. We obtain the number of adults by taking the ratio $\frac{83}{10}$, or 8.3, and then rounding up to the nearest whole number. Alternatively, we can simply round up the number of children to the nearest 10 and then 'ignore the zero'. Now let us assume that if the children from school A go on an outing 12 adults will be required; and if the children from school B go on the outing 14 adults would be required. How many adults would be required if schools A and B combined for the outing? To obtain the answer we reason as follows. We know that there are between 111 and 120 children from school A, and between 131 and 140 children from school B going on the outing. Thus, if the schools combine, between 242 and 260 children will go on the outing. Thus we may need 25 or 26 adults. Which number is more likely? It turns out that the probability we would need 25 adults is $\frac{9}{20}$ (= .45) and the probability that we need 26 adults is $\frac{11}{20}$ (= .55). This means that we would expect to save an adult's time (and the school district's money) 9 times out of 20 by combining the schools when going on outings.

From the point of view of the arithmetic, then, what are the factors involved in deciding how to do an addition or subtraction problem involving approximations? *First,* we must realize that, for problems involving approximations, we have different rules for addition problems and subtraction problems. *Second,* we must take account of whether the data is presented to us rounded up, rounded down, or rounded off. *Third,* we must take account of whether our answer is to be rounded up, rounded down, or rounded off. Having recognized these aspects we begin to realize that the arithmetic of approximations is rather subtle, and we will have to proceed carefully, and systematically, if it is to be done properly.

We will give the rules covering various common situations after discussing a third example. Suppose we are rounding off to the nearest hundredth and have to add the approximations 17.24 and 8.37 (this problem might well arise in adding lengths measured in meters). Then the obvious answer of 25.61 is the most likely, but it is easy to see that the actual answer might be 25.60 or 25.62. For example, the measurements might more accurately have been recorded as 17.236 and 8.366, when the sum would plainly be 25.60 to the nearest hundredth; or they may have been recorded as 17.244 and 8.373, when the sum would be 25.62, to the nearest hundredth.

In giving the rules, we will confine ourselves to those problems that you are most likely to encounter. We will, in accordance with our usual policy, indicate the general principle by means of a typical special case; moreover, we will always round to two places of decimals, trusting to you to make the necessary change in the rules if we are rounding to a different degree of accuracy. If the solution to a problem could be 25.51, 25.52 or 25.53 we will represent this as 25.52 ± 0.01. It should then be understood that 25.52 is the most likely answer.

First we will deal with *addition.* There are five important cases, and they may be classified as follows:

Rules for Adding Approximations

Case	Data rounded	Solution rounded	Sample problem	Solution
1	Up	Off	12.81 + 3.42	16.22 ± 0.01
2	Off	Off	12.81 + 3.42	16.23 ± 0.01
3	Down	Off	12.81 + 3.42	16.24 ± 0.01
4	Up	Up	12.81 + 3.42	16.23 or 16.22
5	Down	Down	12.81 + 3.42	16.23 or 16.24

Next we display the appropriate results for *subtraction*. Here there are seven important cases which we classify as follows:

Rules for Subtracting Approximations†

Case	Data rounded	Solution rounded	Sample problem	Solution
1	Up	Off	9.31 − 6.42	2.89 ± 0.01
2	Off	Off	9.31 − 6.42	2.89 ± 0.01
3	Down	Off	9.31 − 6.42	2.89 ± 0.01
4	Up	Up	9.31 − 6.42	2.89 or 2.90
5	Down	Down	9.31 − 6.42	2.89 or 2.88
6	Up-Down	Up	9.31 − 6.42	2.89 or 2.88
7	Down-Up	Down	9.31 − 6.42	2.89 or 2.90

† In the first five cases (as in addition) we round both numbers in the subtraction problem the same way.

(Mutter, Mutter, Mutter...)

Let us explain, and exemplify, the last two cases, which are special to subtraction. Suppose we know approximately the total length, a, of our journey and the distance we have traveled already, b. There are circumstances in which we might seek a *minimum* value for the remaining distance, $a - b$ (we might want to phone friends to tell them not to expect us before a certain time); and there are circumstances in which we might seek a *maximum* value for the remaining distance, $a - b$ (do we have enough gas to complete the journey?). The minimum problem is case (7); we should round a down and b up. The maximum problem is case (6); we should round a up and b down.

You may be wondering what use you are expected to make of what you are now learning; you may well be muttering rebelliously that we are not true to our principles; that, having said you should not rely on your memories, we are now overloading your memories with strange rules for a strange arithmetic. Actually, our purpose is different; it may be analyzed into the following aspects:

(a)   We want you to realize that handling approximations and estimates is not easy (we didn't find it so ourselves!). It does not follow immediately from the rules of elementary arithmetic.

(b)   We want you to understand that there are many different situations to be dealt with, and that the particular situation depends not on our choice but on the nature of the data and the nature of the problem.

(c)   We want you to understand that the most obvious answer is reasonable but uncertain, and that, in any case, in the arithmetic of approximations, *the answer is less certain and less accurate than the data.*

(d)   We want you to have an accessible source to consult if you need to be precise about the reliability of your answer.

(e)   We want you to have some understanding of how the results presented in these tables are obtained. Thus, in the exercises you will be led by a sequence of questions to some of the results presented in these tables. An understanding of how the results were obtained in a few cases should enable you to work out other cases if you ever have such a need.

Before proceeding to discuss multiplication and division, we should make an important remark. For certain approximate calculations, and most especially when doing an approximate calculation to check an accurate machine calculation, you will not need to bother as to whether you rounded the data up, down or off, nor how you should round the answer. You will be content with a 'rough estimate' and you will not need to know just how accurate your estimate and your answer are. *Nothing we have said in this section applies to such a situation.* You may well find it convenient, in a rough addition of this kind, to round one number up and the other number down, for example—feel free to do so (when you know it makes sense)! We are here really concerned with the situation in which the nature of the problem, and of the measuring instruments used, naturally determines how the data is rounded and how the solution should be rounded.

We turn next to multiplication. To develop a precise set of rules as we have given for subtraction would involve us (and you!) in excessive complication which would not be justified by its practical advantages. Moreover it is to be expected that most multiplications will be done on a hand calculator, so that it will largely be a matter of having a reasonable sense of the degree of accuracy and reliability of our answers. Thus we concentrate on the technique of approximate multiplication and the reliability of the answer.

The precise product of 21.2 and 4.7 is 99.64. However if we are told that a plot of land is 21.2 m by 4.7 m it would be absurd to say that its area is 99.64 m² (square meters), since we obviously can put no faith in the digit 4 in the second place of decimals. But, in fact, the answer is less reliable than even that remark suggests. For if the plot had measured 21.15 m by 4.65 m, its area would be about 98.3 m² (since the precise calculation yields 98.3475); and if it had measured 21.25 m by 4.75 m, its area would be about 100.9 m² (the precise calculation giving 100.9375). Thus the figure we have obtained (rounding off to the nearest tenth) of 99.6 m² is the value of the midpoint between the two extreme possibilities (the average). Moreover, the spread, around the value 99.6 m², of possible values is ±1.3 m², and 1.3 is

$$.05 \times (21.2 + 4.7),$$

where we take the possible error in our measurements of length (.05, since we are measuring to the nearest tenth) and multiply by the sum of the lengths).

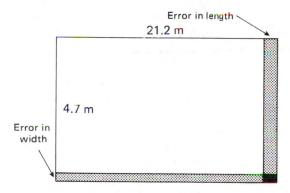

Visualizing the error in an area calculation

This is a very good approximate rule, whose validity we will establish in the final section of this chapter. If we wish to multiply two numbers, each rounded off to a given degree of accuracy, we multiply on the hand calculator and round off to the same degree of accuracy. This is then our *best estimate* but it is subject to an error of

(3.1) ± (maximum error in our measurements) × (sum of the numbers).

Let us give another example of this rule. If the density of a substance is 9.11 (mass per unit volume) and if the volume is 12.35 (measured in some appropriate unit, say cm³), then its mass (measured in appropriate units) is approximately 9.11 × 12.35. From the hand calculator we obtain the estimate 112.51. The possible error, calculated from (3.1) is

$$\pm(.005) \times (9.11 + 12.35) = \pm(.005) \times (21.46) = \pm 0.11.$$

Very frequently estimates to be multiplied will be presented to us in scientific notation. Thus, we may be required to find the product

$$(3.1 \times 10^5) \times (4.6 \times 10^2).$$

We proceed as above to find the product 3.1×4.6, which is 14.3 ± 0.4. Then we use the law of exponents to complete the multiplication. Thus, since $10^5 \times 10^2 = 10^{5+2} = 10^7$, we obtain the final answer

$$(3.1 \times 10^5) \times (4.6 \times 10^2) = 14.3 \times 10^7.$$

There are two important things to notice about this answer. The first is that the possible error of $\pm.04$, in the calculation that gave us 14.3, should also be multiplied by 10^7 to give the possible error in our answer. Thus our answer may be larger or smaller than 14.3×10^7 by an amount of $.04 \times 10^7$, or 400,000. The second thing to notice is that the answer given is not, in fact, in scientific notation, since 14.3 does not lie between 1 and 10. Thus, to put the answer in scientific notation we should rewrite it as

$$1.43 \times 10^8.$$

We saw in the previous section that the use of scientific notation could be extended to numbers less than 1 by using negative exponents. As to be explained in Section 2 of Chapter 5, the law of exponents continues to hold in this extended sense, provided we interpret addition involving negative numbers in the following way, rendered reasonable by considering the number line extended in both directions.

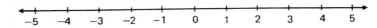

Thus

$$(-2) + (-4) = -6,$$

$$(-3) + 7 \quad = 4,$$

$$(-5) + 2 \quad = -3.$$

With the rules that these examples represent for adding negative numbers we can continue to multiply in scientific notation as above; note however, that our rule requires us to regard 10^0 as 1. Thus

$$(1.25 \times 10^2) \times (.720 \times 10^{-2}) = 9.00 \times 10^0 = 9.00.$$

We now make two important remarks.

Remarks

(i) In addition and subtraction situations, the two quantities being added or subtracted are always of the *same kind*. We add lengths, we subtract sums of money; we do not add a length to a weight, nor do we subtract an area from a volume. Thus it is natural that, in such situations, the two numbers involved be given to the same level of accuracy. However, in multiplication this is often not so. We may indeed multiply quantities of the same kind—as, for example, when we multiply lengths to obtain areas. But very often we have to multiply numbers representing very different quantities. Thus we may multiply speed by time to obtain distance traveled; we may multiply unit cost by quantity purchased to obtain total cost; we may multiply mass by acceleration to calculate force. In these cases there is, of course, no reason why the measurements made should be expressed by decimal numbers recorded to the same degree of accuracy—it is, of course, quite meaningless to ask for the *same* degree of accuracy when *different* measures are involved. There are rules governing the uncertainty involved when this situation arises, but they are more complicated than those given and we are not going to insist on them here (they will, in fact, appear in Volume 3). As a practical rule we recommend that, if a real-life problem gives rise to a multiplication of two approximate measures, you should round off your answer (obtained on the hand calculator, probably) to the lesser accuracy of the two numbers involved, and simply announce that as your best estimate. Thus purchasing 1260 articles (to the nearest 10) at $0.314 each (the unit price being given to the nearest thousandth of a dollar, or *mil*), then we find $1260 \times 0.314 = 395.640$, so the total cost should be announced as $400, as closely as one can say.

(ii) As a variant of the situation discussed in Remark (i) (some of you may prefer to regard it not as a variant but as a special case— that is the true mathematical spirit!) there are many situations in which we have to multiply an estimate or an approximate amount by an *exact* whole number. We may, in other words, be using multiplication by a whole number as repeated addition, and we will often then know precisely how many times to take our estimated amount. If the time of sunset advances 600 seconds each day in May it will advance $7 \times 600 = 4,200$ seconds in a week and we are not going to approximate to 7 to do this calculation! How then should we handle this type of question? Here the exact calculation gives the best estimate—that is clear; but how far out might we be? The answer is very simple. If we are multiplying X by N, where X is a measure to the nearest hundredth, say, and N is an exact whole number, then our best estimate is $N \times X$ and our possible error is $\pm \frac{N}{200}$. We will illustrate this in the exercises.

We do not wish to say too much about division here, since division was treated very extensively in the previous chapter. We can give a reasonable rule for dividing one approximation by another. If both numbers are given to the same degree of accuracy—this presupposes that both numbers represent measurements of the same kind,

so that their ratio is a *number* (not a length, a weight, a time interval, and so on), then we may divide one by the other on our hand calculator and round off our answer either (a) to the nearest whole number if the nature of the problem forces the answer to be a whole number or (b) to the same degree of accuracy as our original measurements if a decimal answer is sensible.† Of course, it is understood in alternative (a) above that our original measurements were themselves at least as accurate as to the nearest whole number. Let us illustrate what we have said so far with two examples, before proceeding to discuss the question of possible error.

If we know that the weight of an individual tack is 23 grams and if we have a quantity of tacks weighing 20120 grams, then our best estimate for the number of tacks is 875; it would be absurd to announce an answer of 874.8, since we must have a whole number of tacks! If on the other hand, we are told that the atomic weights in appropriate units (the weight of a hydrogen atom, say), of two chemical elements are 94 and 91 then the ratio of their weights may be given as 1.03.

Now for the possible error. If we have to divide X by Y and each of these quantities is given, say, to three places of decimals, so that the possible errors in X and Y are $\pm h$, where $h = 0.0005$, then, as we show in the final section of this chapter, the possible error in $\frac{X}{Y}$ is

(3.2) $$\pm \frac{h \times (X + Y)}{Y^2} .$$

Note that this error may be dangerously large if Y is very small compared with X—and this is a rather common situation.

---

†We are never justified in giving an answer to such a division problem to more significant figures than are given in the divisor.

For example, if our measurements of weight above were accurate to the nearest gram, then $X = 20120$, $Y = 23$, $h = 0.5$, so that, using (3.2), the possible error is

$$\frac{\pm 0.5 \times (20120 + 23)}{23^2}$$

$$= \pm \frac{0.5 \times 20143}{529}$$

$$= \pm 19.04, \text{ approximately.}$$

If we are dealing with a division situation in which X and Y represent quantities of a quite different kind (for example, distance and time)—so that their ratio is not a 'pure number' as in the case of the weights above—then the best practical rule we can give is to accept no answer containing more significant figures than there are in the divisor. (Of course, we may be required, in the statement of the problem, only to make a crude estimate of the quotient, but we are here assuming that it is up to us to choose the appropriate level of accuracy.) A warning is necessary here, however, as in our second remark relating to multiplication: namely, the divisor may well be an exact and precise whole number. In this case, of course, the practical rule relating to significant figures in the divisor is no help to us and we must be guided by the actual problem itself, the purpose of the calculation and, in the final resort, by the degree of accuracy of the dividend.

All that we have been doing in this section is to suggest some helpful rules. We have very deliberately been quite undogmatic. The strong impression we hope to have left with you is that, when doing a calculation involving approximations and estimates, your 'solution' is itself also an estimate—and, generally speaking, not as reliable an estimate as those that formed the data for the calculation.

Exercise 4.3

1. In the text example about how many adults would be required to accompany children on an outing if two schools combined for the outing, certain probabilities were asserted. You will now see how these probabilities could have been obtained. Below we list the possible number of children from each of the schools who went on the outing. Fill in the table to show how many children altogether would have gone on the outing in each case.

    (a) How many cases are there (that is, how many entries in the table)?

    (b) How many cases produce a total number of children less than or equal to 250?

    (c) How many cases produce a total number of children greater than 250?

(d)   In how many cases would 25 adults have sufficed?

(e)   In how many cases would 26 adults have been needed?

(f)   What is the probability that 25 adults would have sufficed?

(g)   What is the probability that 26 adults would have been needed?

School A \ School B	131	132	133	134	135	136	137	138	139	140
111	242	243								251
112	243								251	
113								251		
114							251			
115						251				
116					251					
117				251						
118			251							
119		251								
120	251									

2.   Assuming the same rules, discuss in a similar way the question of the number of adults required if school A, on its own would have required 7 adults and school B would have required 3.

3.   A change of rules has occurred! Now an adult is needed to accompany 6 children. Discuss, in a similar way to the two questions above, the number of adults required if schools A and B would each have required 5 adults if undertaking the outing on their own.

4.   Suppose we are given two numbers, X and Y, lying between 0 and 1.

(a)   What is the range of possible values for $X + Y$?

(b)   What is the probability that $X + Y$ is less than or equal to 1? (We write this $X + Y \leqslant 1$.)

(c)   What is the probability that $X + Y$ is greater than or equal to 1? (We write this $X + Y \geqslant 1$.)

(d)   What is the probability that $X + Y$ is nearer to 0 than to 2? (Hint: See your previous answers.)

(e)   What is the probability that $X + Y$ is nearer to 2 than 0?

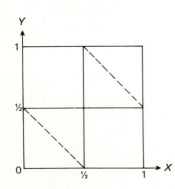

Suppose now that we want to round $X + Y$ off to the nearest whole number. Call this whole number Z.

(f)   What are the possible values of Z?

(g)   Which value do you think is the most likely?

We will now take you carefully through the argument (as promised in the text). Let us divide the range of values of X into two equal parts, 0 to $\frac{1}{2}$

$(0 \leqslant X \leqslant \frac{1}{2})$ and $\frac{1}{2}$ to 1 $(\frac{1}{2} \leqslant X \leqslant 1)$. Call X small if it lies in the first part and large if it lies in the second part. Do the same for Y.

(h)   What is the probability
    (i)   that X and Y are both small,
    (ii)   that one of them is small and the other large,
    (iii)   that X and Y are both large?

(j)   Show that if X and Y are both small then Z is equally likely to be 0 or 1. (Hint: Look carefully at your answer to (d).)

(k)   Show that Z can only be zero if X and Y are both small.

(l)   What is the probability that Z is 0? (Hint: Look at your answers to (h), (j) and (k).)

(m)   What is the probability that Z is 2? (Hint: Look at your answer to (l) and use symmetry.)

(n)   What is the probability that Z is 1?

5.   Suppose two lengths are recorded, rounded off to the nearest tenth of a meter, as 108.5 m and 71.7 m.

(a)   What is the possible range of values for the first length?

(b)   What is the possible range of values for the second length?

(c)   What is the possible range of values for the sum of the lengths?

(d)   What are the possible values for the sum of the lengths rounded off to the nearest tenth of a meter?

(e)   Which is the most likely of these values? What is its probability? What are the probabilities of the less likely values? (Hint: Look at your answers to Problem 4.)

6.   Execute a program like that in Problem 5 for the rounded-off sum of two lengths rounded *down* to 16.37 km and 12.54 km. (Round off in (d)).

7.   Execute a program like that in Problem 5 for the rounded-off sum of two lengths rounded *up* to the nearest millimeter as 7.924 m and 5.381 m.

8.   (a)   Justify the first row of the table giving *Rules for Subtracting Approximations.*

(b)   Give an informal explanation as to why the *solution* is the same for each of the first three rows of the table.

9.
A   B   C

Suppose the length AC, rounded up to the nearest meter, is 12 m and that the length BC, rounded up to the nearest meter, is 5 m. What can we say of the length AB rounded up to the nearest meter? Is there more than one possibility? If so, how many possibilities are there, and which is the most likely?

10.   Repeat Problem 9 replacing 'rounded up' everywhere by 'rounded down'.

11.   In each of the following, the data is given to the nearest hundredth. Estimate the product and give the possible error explicitly.

(a) 5.72×4.13

(b) 12.73×6.27

(c) 10.00×8.31

(d) 0.12×108.46

(e)   Express the problems and the solutions in (a) through (d) in scientific notation.

12.   Estimate *seven* times the following lengths given to two places of decimals, and give the possible error explicitly.

(a)   1074.18 miles

(b)   12.30 km

(c)   341.26 m

13.   If $X = 4.73$ and $Y = 49.87$, estimate $\frac{X}{Y}$ to two places of decimals and give the possible error explicitly, using (3.2).

4.4 GRAPHICAL PRESENTATION OF DATA; AVERAGES AND MEASURES OF SPREAD

In this section we take up rather a different aspect of the analysis of data. We often take measurements in order to get a picture of a process. How is the growth of a plant varying with time? How does the amount of sunlight vary with the seasons? How, quantitatively, does inflation affect prices and consumption? We are generally dealing with two quantities which vary in relation to each other and we want to have some picture—and, thereby, we hope, some understanding—of that relation. Often the data is presented to us in tabulated form, but we may get a better idea of the relationship by some kind of graph. Let us give some examples.

Let the heights in centimeters of 32 boys who were in grade 4 in a certain school in 1978 be as follows.

Boys (in alphabetical order)	Height (in cm)	Boys (in alphabetical order)	Height (in cm)
1	137.4	17	123.9
2	142.7	18	135.8
3	131.6	19	143.6
4	146.2	20	141.2
5	129.1	21	127.9
6	151.2	22	147.8
7	137.1	23	154.9
8	131.6	24	135.2
9	125.8	25	141.5
10	143.9	26	120.4
11	138.2	27	144.7
12	151.7	28	133.2
13	130.4	29	149.6
14	139.7	30	138.9
15	148.2	31	127.9
16	134.7	32	142.7

Figure 1

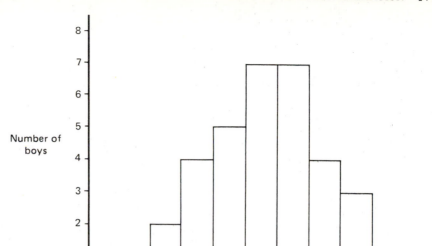

Figure 2

It is extremely difficult to get any picture from this table. Plainly it would be better to list the boys in order of increasing height. But it is even better to present this information in graphical form, as in Figure 2; we call this a *bar graph*. Some information has been lost, but the picture is much clearer.

If we want to compare the heights of the boys with those of girls of the same age in the same school, we can see the comparison more easily and more vividly from the graphs than from the raw data. Here is the corresponding information about the girls.

Girls (in alphabetical order)	Height (in cm)	Girls (in alphabetical order)	Height (in cm)
1	136.7	17	123.4
2	134.7	18	135.1
3	124.8	19	134.1
4	132.4	20	147.9
5	127.2	21	124.8
6	138.2	22	127.2
7	146.1	23	113.4
8	130.2	24	137.4
9	144.5	25	126.1
10	135.1	26	142.1
11	134.2	27	138.9
12	118.2	28	132.6
13	144.2	29	147.2
14	136.0	30	131.4
15	128.9	31	117.1
16	131.7	32	121.2

Figure 3

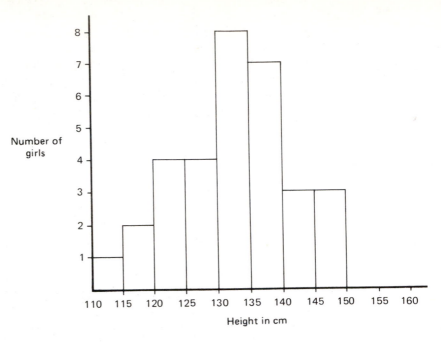

Figure 4

Now suppose we want to study the growth rate of the boys. We follow them through the grades, measuring their heights at six-month intervals up through sixth grade. We have now accumulated a great deal of data, but it is hard to digest and to present. If we are interested in a particular boy, then we may record his rate of growth on a graph as in Figure 5.

Age	Height in cm
10	138.2
$10\frac{1}{2}$	140.1
11	141.9
$11\frac{1}{2}$	144.3
12	149.5

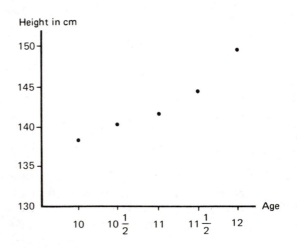

Figure 5

This information, however, is not a reliable indication of the precise nature of the trend toward increased height with age, as we do not know how typical the boy we chose was. A better indication of trend would be provided by seeing how the *average height* varied with time. We would then obtain the graph of Figure 6 (where the averages were taken over a larger sample than that of Figure 1).

Age	Height in cm
10	140.3
$10\frac{1}{2}$	142.3
11	144.2
$11\frac{1}{2}$	146.3
12	148.4

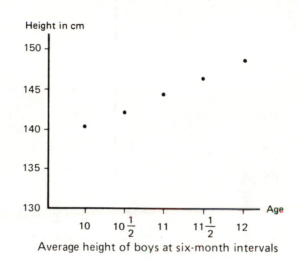

Average height of boys at six-month intervals

Figure 6

A similar situation might be encountered in looking at performances of school children on standardized tests—assuming these unpleasant features of our educational scene continue to plague us for some time, as they probably will. We may choose to follow the progress of certain individual children on a given test and get data like that of Figure 5; or we may wish to compare schools with respect to the performance of their students on such tests and simply record average scores, thus getting data like that of Figure 6. Of course, in making a comparison of this kind between different schools, we must be extremely careful about the conclusions we draw. There may be—indeed, there almost certainly are—differences between the students at two given schools which will strongly influence the academic performances of those students. It follows that differences in average performances on standardized tests should not be regarded as reliable indicators of the quality of the education or of the curriculum. Among the factors which should be taken into consideration are the intellectual capacities of the students (whether due to nature or nurture) and the socio-economic circumstances of their lives.†

†There are, however, fairly sophisticated statistical techniques that enable one to 'factor out' such influences, provided that they have been identified in advance.

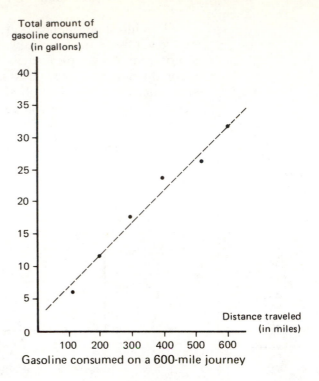

Figure 7

We are not usually content merely to accept a graph such as that of Figure 6—we try to connect up the points in some reasonably smooth way in order to anticipate future trends (this is called *extrapolation*) or, sometimes, to guess intelligently some intermediate reading (this is called *interpolation*). We may, in the first instance, try to put a straight line through the points on our graph. A good example of this is provided by Figure 6; let us look at another example.

It is reasonable, in the case of Figure 7, to draw a straight line more or less as shown. We will not enter here into a description of a mathematical procedure for giving us the best possible fit with a set of given plotted points (the regression line); this can be found, for example, in Moroney's *Facts from Figures* (Pelican Books), a very readable text not requiring advanced mathematics. In practice, we would often be content to make an intelligent guess at the position of such a line.

Sometimes we may notice, from our graph, that certain points 'don't fit'. If we have a reasonable explanation of why those particular readings should be out of step, we may ignore them in looking for patterns. Thus, our figures for average performance of a class on a sequence of tests administered at regular intervals might be the following.

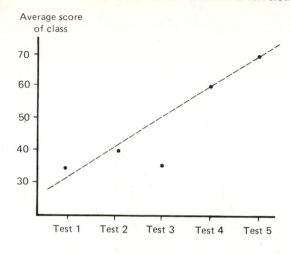

Figure 8

The curious low average for Test 3 could, it turned out, be explained by the fact that a flu epidemic was raging at the time. (Another common explanation of such phenonema of educational statistics is to be found in the scheduling of special TV programs.)

Sometimes—especially when we are concerned with data from a scientific experiment, or series of experiments—we may be satisfied to put a smooth curve through our data instead of a straight line. Thus, for example, the growth rate of population in a certain city was recorded and may be displayed on a graph as shown in Figure 9. In this case, too, our graphical methods enable us to estimate values where no readings have been made or to estimate future values (hence to behave sensibly and rationally in the present). It is for this reason principally that we have included a description of such methods here, in our chapter on estimation.

We have stressed the use of the average (or *mean*) as a way of getting a picture of an aggregate or collection of measurements or readings. But we should also have some means of saying how reliable the average is as a sort of 'representative reading'. Obviously the reliability depends on the spread of the readings around the average (we do *not* say that's all it depends on—we'll revert to this point later). Let us then consider possible measures of spread. One simple measure is the *range* of values—the difference between the highest and lowest reading—but this is clearly very crude since it will depend on just two of the readings; moreover, it has a certain arbitrariness if we retain the right, as suggested, to 'throw out' measurements which don't seem to fit the pattern. A better measure of the spread is given by the average deviation from the mean, provided, of course, that we regard each deviation as positive. Thus, given the readings 2.7, 3.1, 2.9, 3.0, 2.7, 3.0, we compute the average to be 2.9 and the

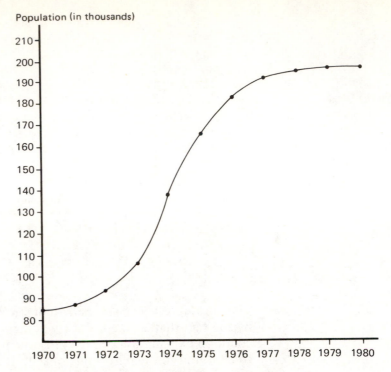

Population of Ecksville in January of each year

Figure 9

deviations are 0.2, 0.2, 0, 0.1, 0.2, 0.1, giving us an average deviation of 0.13. Since this average deviation is small compared with 2.9, we can say in this case that the average, or mean, is a reliable indicator (or *statistic*) for our sample of readings.

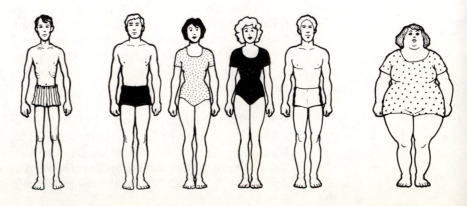

It turns out, however, that for purely theoretical reasons into which we will not enter, it is better to use the squares of the deviations instead of the deviations themselves.† The average of the squares of the deviations is called the *variance* and the square root of the variance is called the *standard deviation*. Thus, in our example, the variance is 0.023 and the standard deviation is 0.15 approximately. Note that if our readings had been measurements of length in meters, then the variance would be in square meters and the standard deviation would again have been in meters.

*

There is a nice theoretical result which helps us to calculate the variance (even when we use a hand calculator!) If we took N readings and our readings are $X_1, X_2, X_3, \cdots, X_N$, then the *mean* or average, M, is given by the formula

$$M = \frac{1}{N} (X_1 + X_2 + X_3 + \cdots + X_N).$$

The squares of the deviations from the mean are $(X_1 - M)^2$, $(X_2 - M)^2, \cdots, (X_N - M)^2$, so that the variance, V, is given by the formula

$$V = \frac{1}{N} ((X_1 - M)^2 + (X_2 - M)^2 + \cdots + (X_N - M)^2).$$

We can, however, often save ourselves a great deal of calculation by observing that, in fact,

Theorem
$$V = \frac{1}{N} (X_1{}^2 + X_2{}^2 + \cdots + X_N{}^2) - M^2.$$

For

$$
\begin{aligned}
(X_1 - M)^2 + \cdots + (X_N - M)^2 &= X_1{}^2 + X_2{}^2 + \cdots + X_N{}^2 \\
&\quad - 2M(X_1 + X_2 + \cdots + X_N) \\
&\quad + NM^2 \\
&= X_1{}^2 + X_2{}^2 + \cdots + X_N{}^2 \\
&\quad - 2NM^2 + NM^2 \\
&= X_1{}^2 + X_2{}^2 + \cdots + X_N{}^2 \\
&\quad - NM^2,
\end{aligned}
$$

(since $X_1 + X_2 + \cdots + X_N = NM$)

---

†This preference has to do with the question of how accurately and reliably we can derive information about a (statistical) population by taking samples of that population.

and the theorem follows immediately. Note that the original formula for V involved N subtractions, N squarings and one division; whereas the theorem gives us a formula for V involving 1 subtraction, $(N + 1)$ squarings and one division. (It is, moreover, an advantage that, if we use the theorem, we square our original readings $X_1, X_2, \cdots, X_N$, rather than the deviations.)

Let us go back to our earlier example to illustrate the theorem. Our readings were 2.7, 3.1, 2.9, 3.0, 2.7, 3.0, so that $N = 6$ and $M = 2.9$. To compute V by the original definition gives us

$$V = \frac{1}{6}\left((0.2)^2 + (0.2)^2 + 0^2 + (0.1)^2 + (0.2)^2 + (0.1)^2\right)$$

$$= \frac{1}{6}\left(0.04 + 0.04 + 0 + 0.01 + 0.04 + 0.01\right)$$

$$= \frac{1}{6}(0.14).$$

If we compute V from the theorem we find

$$V = \frac{1}{6}\left((2.7)^2 + (3.1)^2 + (2.9)^2 + (3.0)^2 + (2.7)^2 + (3.0)^2\right)$$

$$- (2.9)^2$$

$$= \frac{1}{6}\left(7.29 + 9.61 + 8.41 + 9.00 + 7.29 + 9.00\right) - 8.41$$

$$= \frac{1}{6}(50.60) - 8.41.$$

It is easy to check that these two values of V are exactly the same. Of course, this is not an example where the theorem simplifies the calculation of V, since we can calculate V *by hand* very easily from its original definition. The gain in using the theorem will be the greater when (i) N is large and (ii) the spread of readings about the mean is broad.

The incidence of a rare disease!

We close this section with a remark about the use of averages. The average is a useful way of summing up the information contained in a series of readings or measurements, provided we believe that the readings are, in some sense, closely related. Of course, there is always some loss of information in simply reporting the average, but it will not be serious if we are concerned with measurements which are closely related. For example, it was standard practice in the two world wars of this century to take the angle of elevation of a target for one's heavy artillery 6 times (using a device called a director) and then take the average; this was the reading given to the gunners. However, averages can be meaningless and very misleading; we once read an article, written by a medical practitioner nearly 150 years ago, in which he announced that he had determined two authenticated cases of a rare disease, one in a new-born infant, the other in an old woman of 80, and he concluded that the most dangerous age for the onset of the disease was 40! This extreme example illustrates well that it is a theoretical and practical error to average readings which are in no sense taken from the same 'statistical population'.

There is a somewhat more subtle situation in which an average is misleading, even though the readings are related. Suppose we want to estimate what the temperature at noon is going to be in Times Square, New York City, during the month of July, 1982. We take temperature readings each Monday at noon throughout April, May and June. We get 13 readings in this way, we average them, and announce the average as our best estimate. You would surely be ready with an obvious criticism of our procedure. It gets warmer as we proceed from spring through the summer, so that the average reading for April, May and June is a very biased estimate—a considerable *underestimate*—of the expected temperature in July. We have taken no account of the *trend,* or *secular variation* (as many statisticians express it), in temperature when related to seasons of the year.

It is remarkable that this point, so obvious if we are talking of temperatures as they vary through the months of the year, tends to be completely ignored in the standard procedure for measuring a student's performance during the semester! If a student successively scores 70, 80, 90 and 100 percent on the four tests administered in his math course during the fall semester, what significance has the average 85%? Very little, we would say. It is a reasonable measure of his performance level halfway through the semester—that is all. The student is obviously improving steadily and it is absurd, and unfair, to penalize the student just because his climb to a perfect test paper proceeded along a steep path! The moral of the existing grading system is that improvement should not be too rapid! The true moral is that averages can be useful, but they can be meaningless—and dangerous.

Exercise 4.4

1. Compute the average of the heights in Figure 1. How would you set about arranging the heights in increasing order?

2. (a) Deduce from Figure 2 the number of boys in the sample who had heights of 135 cm or over and the number of boys who had heights of 145 cm or under. How would you obtain this information from Figure 1?

   (b) Is the average height you computed in Exercise 1 a good indicator of the heights of these 32 boys?

3. What property or properties of the boys in our sample do you think might be related to their heights?

4. (a) Compute the average of the heights in Figure 3.

   (b) Do you think the difference between the average heights of boys and girls (as recorded in Figures 1 and 3) significant?

   (c) At what ages do you think sex is an important factor in relation to height? How would you propose to test your theory?

5. What do you think John's height (see Figure 5) was

   (a) at age $9\frac{1}{2}$?

   (b) at age $10\frac{1}{2}$?

   (c) What do you think his height will be at age $12\frac{1}{2}$?

   (d) How might you account for the sharp increase in his height between the ages of $11\frac{1}{2}$ and 12?

6. What do you think, judging from the data of Figure 6, is the average height of boys at age

   (a) 9?   (d) $12\frac{1}{4}$?

   (b) $9\frac{1}{2}$?   (e) $12\frac{1}{2}$?

   (c) $11\frac{1}{4}$?   (f) 13?

   (g) Which of these six estimates do you think is the least realiable?

7. (a) Does the *actual* graph of gasoline against distance traveled (Figure 7) pass through the origin?

   (b) Why does the straight line drawn in Figure 7 not pass through the origin?

   (c) What factors, apart from distance traveled, affect gasoline consumption?

8. (a) Use Figure 9 to infer the probable size of the population of Ecksville in July of each year from 1969 to 1981.

   (b) Describe circumstances that might account for the population trend exhibited in the graph.

*9. Compute the mean, variance, and standard deviation for

   (a) the first five entries in Figure 1.

   (b) the first ten entries in Figure 1.

(This exercise is designed to show you the expediency of using the formula $V = \dfrac{X_1{}^2 + X_2{}^2 + \cdots + X_N{}^2}{N} - M^2$ for the variance.)

10. (a) What is the average score if two dice are thrown and the sum of the numbers showing up is taken?

(b) What would be the variance of the 36 scores if we rolled two dice 36 times and each sum between 2 and 12 came up the expected proportion of times?

11. Could it be useful, in any circumstances, to compute and record the average rainfall registered at a given place for each hour of a given 24 hour day?

*12. Show that, if

$$V(A) = \frac{1}{N}\left((X_1 - A)^2 + (X_2 - A)^2 + \cdots + (X_N - A)^2\right),$$

then

$$V(A) - V = (M - A)^2,$$

where M is the mean of $X_1, X_2, \cdots, X_N$. Deduce that the minimum value of $V(A)$ is V, achieved when $A = M$. (This means that the mean square deviation of the readings from a given quantity A is smallest when that quantity is the mean of the readings.)

*4.5 MULTIPLYING AND DIVIDING ESTIMATES

We were concerned in Section 4.3 with the question of the reliability of the estimate $A \times B$ of the product of two numbers whose estimated values are A and B. We also raised a similar question related to division.

Let us suppose then that A is subject to an error of $\pm h$ and that B is subject to an error of $\pm k$. We assume that h and k are positive and small compared with A and B. Then the biggest value the product can have is $(A + h) \times (B + k)$, and the smallest value is $(A - h) \times (B - k)$. Now

$$(A + h) \times (B + k) = (A \times B) + (k \times A) + (h \times B) + (h \times k).$$

However $(h \times k)$ can be neglected as being very small compared with $(k \times A)$ and $(h \times B)$, so that the greatest possible value is, effectively,

$$(A \times B) + (k \times A) + (h \times B).$$

Similarly the smallest possible value is effectively

$$(A \times B) - (k \times A) - (h \times B).$$

We conclude that the estimate $A \times B$ is subject to the error $\pm((k \times A) + (h \times B))$.

We were particularly concerned in Section 4.3 with the case $k = h$. Then the error term is $\pm h \times (A + B)$. This is precisely the rule (3.1) of Section 4.3.

We now pass to the division problem. We suppose that X is subject to an error of $\pm h$ and that Y is subject to an error of $\pm k$, and we again assume that h and k are positive and small compared with X and Y. We seek to measure the possible error in the quotient of X by Y. Now the greatest value the quotient could have is $\frac{X + h}{Y - k}$ and

$$\frac{X + h}{Y - k} - \frac{X}{Y} = \frac{kX + hY}{Y(Y - k)} .$$

Now

$$\frac{1}{Y(Y - k)} - \frac{1}{Y^2} = \frac{k}{Y^2(Y - k)} .$$

Thus this difference is small if k is small, so we may say that

$$\frac{X + h}{Y - k} - \frac{X}{Y} \text{ is effectively } \frac{kX + hY}{Y^2} .$$

Similarly the smallest value the quotient could have is

$$\frac{X - h}{Y + k}$$

and

$$\frac{X}{Y} - \frac{X - h}{Y + k} \text{ is effectively } \frac{kX + hY}{Y^2} .$$

We conclude that the estimate $\frac{X}{Y}$ is subject to the error

$$\pm \frac{kX + hY}{Y^2} ,$$

just as we described in formula (3.2) of Section 4.3. Notice that there is another important special case, namely, that in which there is no possible error in Y (typically, Y is a whole number). Then $k = 0$, so that the estimate $\frac{X}{Y}$ is subject to the error $\pm \frac{h}{Y}$. Of course, this case is particularly easy and there was no need for the elaborate mathematical reasoning above to establish it. We simply wanted to show you that our general argument produced a result that included this easy, but important, special case.

Exercise 4.5

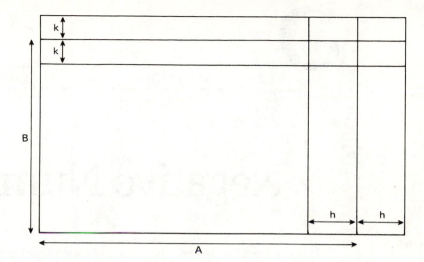

$\left\{\begin{array}{l} \\ \\ \end{array}\right.$ ***1.**   Illustrate the argument relating to *multiplying* estimates (given in this
section) by means of the diagram above.

5

Negative Numbers

We will first be concerned principally with *negative integers*. The last section of the chapter will, however, consider the question of how, and when, to do arithmetic with negative decimals or with negative fractions.

5.1 NEGATIVE NUMBERS; ADDITION AND SUBTRACTION

When we are *counting* objects, there is an absolute quality to the result. We count the apples on the dish and announce there are three; a child counts the fingers on her hand and announces there are five. If somebody else counts the fingers on the child's hand and announces there are six, then it may be stated categorically that that person is wrong.

There is no such absolute quality to measurement—the numbers we use to express the results of measurement depend on the unit we use. Thus we may measure length in inches, or in yards, or in centimeters, or in meters; or, perhaps, in some combination of units. On the other hand, although the measures of length (or weight, or area, or volume, etc.) depend on the choice of unit, there is still, in these cases an absolute meaning to 'zero measure'—weightlessness is weightlessness, whether measured in pounds or grams.

However, there are some measurements in which the 'zero' is also fixed arbitrarily. A very common example is temperature. It is well-known that 0° Celsius is the same as 32° Fahrenheit; these represent the temperature at which water freezes. Another example where we fix an arbitrary zero is with regard to longitude, where the zero is fixed at the Greenwich meridian.

For such measurements it is convenient to have a way to describe measures that are less than zero. As any resident of Chicago knows, there can be temperatures below 0° Celsius; and there can certainly be locations both east and west of the Greenwich meridian. We use *negative numbers* for this purpose; rather, we prefer to say that we *can* use negative numbers for this purpose and it is often convenient to do so. The convenience comes from the fact that we can then continue to use our arithmetic to study situations in which measures below zero occur. But it should be understood that negative numbers are strictly an invention of the human race! A world occupied by distinct objects may be said to contain, implicitly, the notion of the natural numbers 0, 1, 2, 3, · · ·; in no sense can it be held to contain the notion of negative numbers! A 'celebrated' text contains the phrase 'Negative numbers are the degrees below zero on a thermometer'. This is nonsense—the number of degrees below zero in a temperature announced as 'minus 10 degrees' is 10 degrees! But it is *convenient* to replace the phrase '10 degrees below zero' by 'minus 10 degrees' so that to every temperature there should correspond a number; we achieve this effect by enlarging our notion of number to include *negative* numbers as well as our original (*positive*) numbers. Such an enlargement pays off if we can also extend our arithmetic to our enlarged number domain in such a way as to answer questions which naturally arise about temperature, longitude, bank balance, etc.

What might such questions be? Let us take two examples drawn from the study of temperatures.

Example 1 The temperature at midnight was minus 10 degrees Celsius.† By midday it had risen to 12 degrees Celsius. By how much had it risen? The answer is clearly (we hope!) 22 degrees. This can easily be seen—if the reader finds such a device necessary—by constructing a temperature scale in Celsius degrees and measuring the difference. Now differences between numbers are surely obtained by *subtraction*. Thus our arithmetic should yield the fact

$$12 - (\text{minus } 10) = 22.$$

Example 2 The temperature at midday is 12 degrees Celsius. It drops 20 degrees between midday and midnight. What is the temperature at midnight? Again it should be clear that the answer is minus 8 degrees. So our arithmetic should yield

$$12 - 20 = \text{minus } 8.$$

From the world of bank balances we can also draw some examples; here we use negative numbers to represent overdrafts—

---

†Some prefer saying 'negative 10' to 'minus 10'. We will use the latter term.

thus an overdraft of $25 will be called a balance of 'minus $25'. With this interpretation the reader should have no difficulty making up bank balance problems yielding the arithmetical facts

minus 20 + 20 = 0,
minus 15 + 10 = minus 5,
minus 7 + 20 = 13;

of course there are also temperature problems giving rise to the same facts.

Let us grant, then, that it would be useful to be able to add and subtract using negative numbers as well as positive numbers. What should our rules be for such addition and subtraction? If we envisage a scale with a zero mark, positive numbers to the right of the zero, negative numbers to the left,

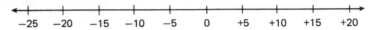

then our examples suggest the following rules:

To add a positive number N, move N places to the right.
To add a negative number minus N, move N places to the left.
To subtract a positive number N, move N places to the left.
To subtract a negative number minus N, move N places to the right.

We may express the second and fourth rules above in the following ways

$$M + (\text{minus } N) = M - N$$

(1.1)

$$M - (\text{minus } N) = M + N$$

Notice that here M is *positive or negative.* The first rule in (1.1), which says that adding minus N is the same as subtracting N, leads us to write '$-N$' instead of 'minus N.' Then (1.1) becomes

$$M + (-N) = M - N$$

(1.2)

$$M - (-N) = M + N,$$

where M is positive or negative, but N is positive. Let us take one further notational step. We know that if N is positive then $-N$ is the corresponding negative number. Thus preceding N by the minus sign has the effect of flipping over from the right to the left of our scale, that is, of 'reflecting in zero'. Why not agree that this works also if N is itself negative? Why not adopt the convention that preceding with the minus sign always has the effect of reflecting in zero? Thus, if A is positive, then $-A$ is the negative number the same distance from 0, and if B is negative, then $-B$ is the positive number the same distance from 0.

What is the advantage of this convention? It is that the two rules (1.2) now become a single rule

(1.3) $\qquad M + (-N) = M - N,$

where *both M and N are now positive or negative (or zero)*. As an example suppose N is negative, say $N = -5$, then $-N = 5$ so that (1.3) asserts, in this case, that

$$M + (5) = M - (-5)$$

which, because 5 is positive, is representative of the second rule of (1.2) (written with the right and left hand sides reversed). We can also notice from this equation that

$$+5 = -(-5).$$

This last example works in general, for if N is negative, say $N = -A$, with A positive, than $-N = A$, so that (1.3) asserts in this case that, with A positive, $M + A = M - (-A)$, and this is exactly the second rule of (1.2). Notice that our convention may be expressed in the neat form

(1.4) $\qquad -(-N) = N.$

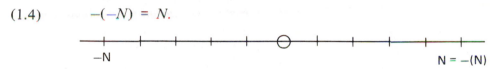

In words this means that 'subtracting $-N$ is the same as adding N'.

Our rule (1.3) (or the two rules (1.2)) tells us how to add and subtract negative numbers; better still, they tell us how to add and subtract *integers,* where an integer may be positive (that is, a natural number), negative (that is, the negative of a natural number), or zero†. We cannot emphasize too strongly that *we have invented the rule to be useful.* (First we invented negative numbers, then we invented the rules for adding and subtracting them.) Had we found no use for adding or subtracting negative numbers we simply wouldn't have defined these operations. There is no way in which we could logically have *inferred* how to add and subtract negative numbers from our knowledge of the arithmetic of positive numbers. We have *chosen* a definition for the addition and subtraction of integers which is (a) useful and (b) consistent with our rules for adding and subtracting positive integers.

†It is a matter of convention, not often adopted at this level, to include zero as a positive integer so that we avoid the awkward term 'non-negative'. The usual convention is to include it as a natural number but to regard it as neither positive nor negative. In this convention it is customary to talk of 'non-negative integers' if one wishes to consider positive integers *and* zero.

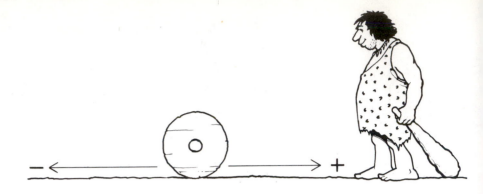

Care should be exercised in presenting answers to problems involving negative numbers. If we are told that, starting 10 miles east of Denver, we travel 12 miles west and are asked what is our new position relative to Denver, we may calculate $10 - 12 = -2$ and announce the answer '−2 miles east of Denver'. Anybody knowing our conventions would interpret our answer as meaning '2 miles west of Denver', so we would have talked good sense, even though it would have been preferable to have given the answer in this latter form. Similarly some data about our financial transactions may lead us to the conclusion that we finish with a bank balance of −$100, and this would be sensible if we interpret this as an overdraft of $100. But if we do a calculation to answer some question and announce that the result is '−6 girls' or '−3 apples', then we are *talking nonsense*; either the question was absurd or we have made a grave error in our calculation (or, just conceivably, we have misinterpreted our mathematics). It is necessary to stress this because, over and over again, students happily produce such answers without being in any way troubled by their obvious absurdity. The absurdity, as we have said, may conceivably have lain in the original question ('4 girls enter an empty room and then 10 girls leave'), in which case the arithmetic tells us the question was absurd—if it was not already obvious. But, if the question was a reasonable one, then an absurd answer must simply be WRONG.

We close with some worked problems to show the reader how to add and subtract integers in practice. The reader is advised to study these examples carefully and to notice that

(a)   in each case we are dealing with just two numbers; and

(b)   the first step in the solution is to write those two numbers so that each appears with exactly one sign (either a '+' or a '−') in front of it. Examples 3 and 5 are already seen to be in this form if we remember that by convention '18' also means '+18' (it is customary to omit the '+' before a number if no ambiguity results).

After step (b) has been achieved you will always be faced with combining two numbers, each of which has one sign in front of it. Now either both numbers will have the *same* sign or they will have *different* signs. Pay particular attention to how the final answer is obtained in each case and see if you can formulate a rule that will help you work such problems. Also use a number line marked off with positive and negative integers if *that* will help (see Figure 1).

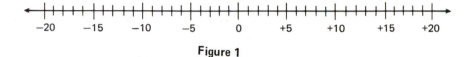

Figure 1

Example 3 $18 - 74$. Here we are subtracting a bigger number from a smaller one. Following the rules developed earlier in this chapter, you will see that what we would actually do is: compute $74 - 18$ (since it's the only way we know how to subtract), getting 56. Then, since the larger number in our problem was negative we take the negative of 56, getting the final answer -56. So, $18 - 74 = -56$. (We have to move 74 places to the left from $+18$. This brings us to a point 56 places to the *left* of 0.)

Example 4 $21 -(-14)$. Subtracting -14 is adding 14, so $21 -(-14) = 21 + 14 = 35$.

Example 5 $-12 -41$. We can interpret this as *adding* 'minus 12' and 'minus 41'. Thus we obtain 'minus 53' or -53. So, $-12 -41 = -53$. (You go 41 steps to the left from -12; you reach a point 53 steps to the left of 0.)

Example 6 $-27 -(-82)$. As in Example 4, we interpret subtracting -82 as adding 82, so we must calculate $-27 + 82$. This is the same as $+82 - 27$ or $82 - 27$. So, as in Example 4 we subtract, getting 55. And, since the larger number was positive our answer is $+55$, or, omitting the '+' sign, 55. So, $-27 -(-82) = -27 + 82 = 82 - 27 = 55$.

Example 7 $-42 - (-17)$. As in Example 6, we interpret this as $-42 + 17$, or $17 - 42$, and now work the example as Example 3, getting the answer -25. So, $-42 -(-17) = -42 + 17 = 17 - 42 = -25$.

Example 8 $83 + (-19)$. This is just $83 - 19 = 64$. So, $83 + (-19) = 83 - 19 = 64$.

Example 9 $17 + (-41)$. This is just $17 - 41$; working as in Example 3, we get the answer -24. So, $17 +(-41) = 17 - 41 = -24$.

Example 10 $-18 + (-37)$. As in Example 5, this is just -55. So, $-18 + (-37) = -18 - 37 = -55$.

Normally, when working this type of problem you would not write out the words that appear in these examples (they were given here by way of explanation). All that you would ordinarily write, in each case, is the last sequence of equalities. However, at this stage, you may find it useful to interpret these calculations as we have interpreted Examples 3 and 5.

Exercise 5.1

1. From your observations of Examples 3 through 10 in this section, complete the following two-part rule for combining two signed numbers:

    (a)   When the two signs are alike you _____ the numbers and put the common sign before the result to obtain the answer.

    (b)   When the two signs are different you _____ the numbers and put the sign of the _____ number before the result to obtain the answer.

2. Write the following expressions as a single integer.

    (a) $5 + 7$ (g) $56 + (-43)$

    (b) $5 + (-3)$ (h) $63 + (-78)$

    (c) $7 - 3$ (i) $50 - 39$

    (d) $-11 + (-16)$ (j) $64 - 85$

    (e) $-7 - (-41)$ (k) $42 - (-32)$

    (f) $45 + 90$ (l) $29 - (-85)$

3. Look in current publications (newspapers, magazines, etc.) to find places where negative numbers are used.

5.2   POSITIVE AND NEGATIVE EXPONENTS

Let us recall the use of the positive integers as *exponents* (or *indices*). If X is any (positive) number then we define X^n, $n = 1, 2, 3, \cdots$, by the rule

$$X^1 = X$$

$$X^2 = X \times X$$

$$X^3 = X \times X \times X$$

$$\cdots$$

$$X^n = \underbrace{X \times X \times X \times \cdots \times X}$$

X appears n times
as a factor.

These statements may be read in various ways. For example, we call X^1 just 'X', the 'first power of X', or (less frequently) 'X to the exponent 1'; we call X^2 the 'square of X', 'X squared', the 'second power of X', or 'X to the exponent 2'; we call X^3 the 'cube of X',

x

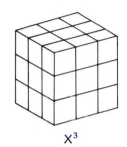

X^2

X^3

'X cubed', the 'third power of X', or 'X to the exponent 3'. In these three cases the last two formulations are especially useful in that we can generalize them and call X^n the 'n^{th} power of X' or 'X to the exponent n'. It is very common, however, to say, briefly, 'X to the nth'. We also may say that, in X^n, n is the† 'exponent' or 'index'. Then we have the following important rules, called the *laws of exponents*:

(2.1) $$X^m \times X^n = X^{m+n},$$

(2.2) $$(X^m)^n = X^{m \times n}.$$

We can see that these rules are reasonable by replacing the letters by particular numbers. Thus, for example if $X = 3$, $m = 2$, $n = 4$, then (2.1) says that $3^2 \times 3^4 = 3^{2+4} = 3^6$; while (2.2) says that $(3^2)^4 = 3^{2 \times 4} = 3^8$. We can verify these from the meaning given above for X^n, but we can, if we please, check directly that the first asserts that $9 \times 81 = 729$, while the second asserts that $9 \times 9 \times 9 \times 9 = 6561$. You may wish to check other cases for special values of X, m and n.

In the expression X^n we are only allowed, by our definition, to take n to be a positive integer. Now, however, that we also have negative integers, it is reasonable to ask whether we could assign any useful meaning to X^n if n is *any* integer, positive, negative, or zero. The key to the answer to this question is contained in rule (2.1); for, if we *can* assign a useful meaning to X^n with n negative or zero, then we would certainly want this rule to remain true, since it is the basis of the effective use of exponents (those familiar with logarithms will well understand this remark). Moreover, we know what we want (2.1) to say, once we have given a meaning to X^n, since we know, from our work in the previous section, what we mean by $m + n$, when m, n are *any* integers.

Let us begin by trying to decide what we should mean by X^0. If (2.1) is to hold for all integers m, n, let us put $m = 0$, $n = 1$. Then it claims that

$$X^0 \times X = X^0 \times X^1 = X^{0+1} = X^1 = X.$$

We observe that when we follow the law (2.1) then multiplying any X by X^0 leaves X unchanged. And the only number we know of that has this property is the number 1. Thus the only meaning we can give to X^0, to maintain the truth of (2.1), is that $X^0 = 1$, whatever value X may have. Let us record this:

(2.3) $$X^0 = 1.$$

†We prefer to use the term 'exponent', since the word 'index' occurs frequently in applied arithmetic—'retail price index', 'sales index', 'cost of living index'—in a totally different sense from that being described here.

Now we want to give a value, say, to X^{-3}, in such a way that the law of exponents (2.1) holds. We put $m = -3$, $n = 3$ into (2.1) and, remembering (2.3), we get

$$X^{-3} \times X^3 = X^{-3+3} = X^0 = 1.$$

This means that X^{-3} must be a number such that when you multiply X^3 by it you get 1. Now recall that $\frac{1}{2}$ is what you multiply 2 by to get 1; $\frac{1}{3}$ is what you multiply 3 by to get 1; $\frac{1}{4}$ is what you multiply 4 by to get 1; etc. So the 'fraction' $\frac{1}{X^3}$ is what you multiply X^3 by to get 1. This means that X^{-3} must be the 'fraction' $\frac{1}{X^3}$. More generally, we see that

(2.4) $X^{-n} = \dfrac{1}{X^n}$, if n is any positive integer.

We have thus given meaning to X^n for *all* integers n, and it is not difficult to verify that we have indeed maintained the truth of (2.1); for example, we should have $X^{-3} \times X^5 = X^{-3+5} = X^2$, and this is true, since

$$X^{-3} \times X^5 = \frac{1}{X^3} \times X^5 = \frac{X^5}{X^3} = \frac{X \times X \times X \times X \times X}{X \times X \times X}$$

$$= X \times X = X^2.$$

The reader should verify some cases of (2.1), involving negative exponents, for herself or himself. There is also a very important remark to be made in connection with (2.4), which asserts that $X^{-n} = \frac{1}{X^n}$ for any positive integer n. We claim that this equality holds for any *negative* integer, too (of course it holds if $n = 0$—why?); provided we adopt the convention (1.4) that $-(-n) = n$. As an example notice that since $X^{-3} \times X^3 = 1$ we have

$$X^3 = \frac{1}{X^{-3}},$$

and, by (1.4) $3 = -(-3)$ so we can write

$$X^{-(-3)} = \frac{1}{X^{-3}},$$

which is just the relation (2.4) with $n = -3$. But there was nothing special about '3' (except that it is a familiar number!) and the argument holds just as well with '3' replaced by 'n'. Thus we may say that the relation (2.4),

$$X^{-n} = \frac{1}{X^n},$$

holds for *all* integers *n*. This is an important mathematical statement; but in practice it is more useful to adjoin to (2.4) the essentially equivalent equality

(2.5)
$$X^n = \frac{1}{X^{-n}} .$$

The use of the rule (2.4) enables us to deduce, in complete generality (i.e. for *any* integers *m, n*) that

(2.6)
$$\frac{X^m}{X^n} = X^{m-n} .$$

We will use (2.6) in future as freely as we use the other rules of exponents. So should you.

Negative exponents do, in fact, give us a real use for negative numbers. Let us think of the metric system. How many meters are there in a kilometer? We can express the answer as 10^3, thus

$$1 \text{ km} = 10^3 \text{ m}.$$

How many kilometers are there in a meter? With the help of negative exponents our answer is just as simple,

$$10^{-3} \text{ km} = 1 \text{ m, or } 1 \text{ m} = 10^{-3} \text{ km}.$$

Similarly, we have†

$1 \text{ m} = 10^2 \text{ cm}$,		$1 \text{ m} = 10^2 \text{ cm}$,
$10^{-2} \text{ m} = 1 \text{ cm}$,	or	$1 \text{ cm} = 10^{-2} \text{ m}$,
$1 \text{ km} = 10^5 \text{ cm}$,		$1 \text{ km} = 10^5 \text{ cm}$,
$10^{-5} \text{ km} = 1 \text{ cm}$,	or	$1 \text{ cm} = 10^{-5} \text{ km}$.

and many other such conversions, expressed in this very economical and natural way, so that we can go 'from left to right' or 'from right to left' with equal facility.

Scientists are very prone to talk of 'orders of magnitude'—an important concept in estimation. Thus, instead of referring to the annual defense budget of the U.S.A. as '120 billion dollars' or '$120,000,000,000', they would be apt to say it is

(2.7)
$$\$1.2 \times 10^{11}.$$

Such a way of writing is, as explained in Section 4.2, called *scientific notation*. Notice that the first number '1.2' in (2.7) has been taken between 1 and 10—this is customary so that comparisons may be

†Both columns of statements state exactly the same facts, but we've changed the order of terms appearing on the two sides of an equality. We've noticed that some students have a definite preference for one or the other—hence we show both. You may use whichever form you find the more intuitive and easier to remember.

made more easily. The use of scientific notation helps us enormously to estimate and to keep significant magnitudes in our heads—words like 'billion' and 'million' are not at all helpful for this purpose and indeed often serve to confuse.†

Now very small numbers can also be expressed vividly using scientific notation provided that we take advantage of our freedom to use negative exponents. For example, the mass of an electron is approximately

(2.8) 9.1×10^{-31} kg

which is, again approximately

(2.9) 5.5×10^{-4} atomic mass units.

The convenience of expressing enormously big numbers like (2.7), or very small numbers like (2.8), (2.9), in scientific notation, employing exponents, should be obvious. Moreover, approximate calculations can then be done very easily.

The reader may be surprised that we have made no mention of the second law of exponents (2.2), beyond just writing it down and verifying it—all our work, thus far, has been with (2.1). Now, however, (2.2) is going to come into its own!

Some of you may have been asked to simplify (say, on a standardized test) an expression that looked something like this:

$$\frac{(A^3 \times A^{-1}) \times B^3}{(A \times B)^{-2}} .$$

†Governments usually spend 'billions' and announce savings of 'millions'—and the voters too often are deceived by this!

This instruction usually meant that you were to write an expression equivalent to this one so that it is a single fraction with no negative exponents. If you had to play that game, you could proceed systematically by writing the expression first as

$$A^3 \times A^{-1} \times B^3 \times \frac{1}{(A \times B)^{-2}} \ .$$

Then, since A^{-1} means $\frac{1}{A}$, and since $\frac{1}{(A \times B)^{-2}}$ means $(A \times B)^2$,

we replace the former by the latter in each case, and obtain

$$A^3 \times \frac{1}{A} \times B^3 \times (A \times B)^2 \ .$$

Now, going back to the definitions of the exponents we can write this as

$$\frac{A \times A \times A \times B \times B \times B \times (A \times B) \times (A \times B)}{A} \ .$$

Then, on dividing out the A in the top and bottom we obtain (ignoring the parentheses),

$$A \times A \times B \times B \times B \times A \times B \times A \times B$$

or, again using the definition of an exponent, we write the final answer as

$$A^4 \times B^5 \ .$$

The above description shows rather explicitly what is happening at each point. It is useful for you to be able to carry out these steps, but as you become more familiar with the concepts you would probably want to shorten what you write down. Your calculations might eventually look something like this:

$$\frac{(A^3 \times A^{-1}) \times B^3}{(A \times B)^{-2}} = \frac{A^3 \times B^3 \times (A \times B)^2}{A^1}$$

$$= \frac{A^3 \times B^3 \times A^2 \times B^2}{A^1}$$

$$= \frac{A^{3+2} \times B^{3+2}}{A^1}$$

$$= \frac{A^5 \times B^5}{A^1}$$

$$= A^{5-1} \times B^5$$

$$= A^4 \times B^5$$

As a matter of fact you would probably not want to write down all of these steps; We included them to help you see where the answer came from. One feature of the calculation above deserves special mention. Notice that we went from $(A \times B)^2$ to $A^2 \times B^2$. It should be clear to you that this step is valid and that, indeed, a more general rule holds, namely,

(2.10) $$(A \times B)^n = A^n \times B^n.$$

This rule combines nicely with (2.2) to give us the rule

(2.11) $$(A^p \times B^q)^n = A^{p \times n} \times B^{q \times n}.$$

We will use this rule in one of our examples.

Calculations like those in our 'standardized test' example above do not often arise in real life (there are important applications to 'dimensional analysis' which you may meet later). However, you *may* be asked to do such a calculation on some test; if so, please look carefully to see if negative exponents are allowed in your final answer. For we claim that, for example, A^{-5} is quite as good an expression as $\frac{1}{A^5}$, and in many respects better. We do not, however, find $\frac{1}{A^{-5}}$ as 'final' a form as A^5 (just as $-(-7)$ is better replaced by 7).

Here are two more examples of this kind of calculation. Again, more steps may have been included than you will wish to include in your own calculations. It may be helpful for you to compare the various options that are illustrated and then you can use the form that is most comfortable for you.

Example 1   Simplify: $A^2 \times A^{-4}$.

Solution (i) $A^2 \times A^{-4} = A^{2+(-4)} = A^{-2}$ or $\frac{1}{A^2}$.

Solution (ii) $A^2 \times A^{-4} = \frac{A^2}{A^4} = \frac{A \times A}{A \times A \times A \times A} = \frac{1}{A^2}$ or A^{-2}.

Solution (iii) $A^2 \times A^{-4} = \frac{A^2}{A^4} = \frac{1}{A^4 \times A^{-2}} = \frac{1}{A^{4+(-2)}} = \frac{1}{A^2}$ or A^{-2}.

Example 2   Simplify: $(A^2 \times B)^3 \times A^{-2} \times B^{-4}$.

Solution (i) $(A^2 \times B)^3 \times A^{-2} \times B^{-4} = \dfrac{(A^2 \times B)^3}{A^2 \times B^4}$

$$= \frac{(A^2 \times B) \times (A^2 \times B) \times (A^2 \times B)}{A^2 \times B^4}$$

$$= \frac{A \times A \times B \times A \times A \times B \times A \times A \times B}{A \times A \times B \times B \times B \times B}$$

$$= \frac{A \times A \times A \times A}{B}$$

$$= \frac{A^4}{B} \quad \text{or} \quad A^4 \times B^{-1}.$$

Solution (ii) $(A^2 \times B)^3 \times A^{-2} \times B^{-4} = \dfrac{A^{2 \times 3} \times B^3}{A^2 \times B^4}$

$$= \frac{A^6 \times B^3}{A^2 \times B^4}$$

$$= \frac{A^{6-2}}{B^{4-3}}$$

$$= \frac{A^4}{B} \quad \text{or} \quad A^4 \times B^{-1}.$$

Obviously, in this example, Solution (ii) is much more efficient than Solution (i). We wish to reemphasize that the steps you will write down in your own solutions should depend on your own familiarity with the concepts. When you become really expert you would simply write down the final answer for each of these examples without *any* intermediate steps.

In both these solutions we've immediately eliminated negative exponents, and only restored them to present the final solution. If you are comfortable with negative exponents, you may, however, prefer to argue as follows.

Solution (iii) $(A^2 \times B)^3 \times A^{-2} \times B^{-4} = (A^6 \times B^3) \times (A^{-2} \times B^{-4})$

$$= (A^6 \times A^{-2}) \times (B^3 \times B^{-4})$$

$$= A^4 \times B^{-1}.$$

Exercise 5.2

1. Why do you suppose some people might think 1 billion dollars sounds like less than 9 million dollars?

2. Since one billion = 1,000,000,000, one billion dollars may be written in scientific notation as 1×10^9 dollars (this would, of course, in practice be abbreviated to 10^9 dollars). Write 9 million dollars in scientific notation, and use this notation to see that 1 billion is much bigger than 9 million.

3. If it costs the local school district $100,000 per year to operate the schools in their district, how many years could they run the schools if they had

   (a)   9 million dollars?

   (b)   1 billion dollars?

   (Don't worry, in this problem, about how much interest might be collected on the unused money before it is spent.)

4. Express 90 years and 10,000 years in scientific notation, and compare the two expressions.

5. When comparing the size of two numbers expressed in scientific notation, say 3×10^{12} and 6×10^2, is it more important to pay attention to the first digit (the 3 and the 6 here) or to the exponents (12 and 2)? Why?

In problems 6 through 16, simplify the given expressions (see Examples 1 and 2 for possible methods).

6. $A^3 \times A^{-2}$

7. $B^5 \times B^{-3}$

8. $\dfrac{B^5}{B^{-2}}$

9. $\dfrac{A^4}{A^{-3}}$

10. $\dfrac{A^2 \times B^{-2}}{B^3}$

11. $A^3 \times (B \times C)^{-1} \times B^4$

12. $(A \times B \times C)^2 \times (A \times B \times C)^{-2}$

13. $(A^2 \times B^3)^3 \times (A^{-2} \times B^{-3})^3$

14. $(A^3 \times B^4)^2 \times (A^4 \times B^3)^{-2}$

15. $(A^{-2} \times B^3)^3 \times (A^3 \times B^{-2})^2$

16. $(A^4 \times B^{-3})^{-2} \times (A^{-2} \times B^{-3})^{-1}$

17. Give examples to show that rule (2.6) fits with (2.3), (2.4) and (2.5).

5.3 MULTIPLICATION OF NEGATIVE NUMBERS

In this section we describe the rule which has proved most useful for multiplying (and dividing) negative numbers, and which is universally adopted. Again we wish to emphasize that this rule has been chosen for its convenience and usefulness—it is not a rule that can be *deduced* from our knowledge of how to multiply positive numbers.

To show how natural the rule is (especially 'negative times negative is positive'), we revert to the law of exponents (2.2), which we now repeat as

(3.1) $(X^m)^n = X^{m \times n}.$

We are now in the position of knowing what we mean by $(X^m)^n$ where m, n are any integers, but we do not know yet what $m \times n$ should mean unless m, n are both positive. Let us then see if we can give a meaning to $m \times n$ in such a way that (3.1) remains true for *any* integers m, n.

Let us first consider $(X^2)^{-3}$. Then, by (2.4), we know that

$$(X^2)^{-3} = \frac{1}{(X^2)^3} = \frac{1}{X^6} = X^{-6}.$$

Thus we see that if (3.1) is to remain true, we should have

$$2 \times (-3) = -6$$

and, since there was nothing special about the numbers 2 and 3, we could repeat the argument substituting '*m*' for '2' and '*n*' for '3' to obtain the more general result

(3.2)
$$m \times (-n) = -(m \times n),$$

for any two positive integers *m, n*.

Next let us consider $(X^{-3})^4$. Then, again using (2.4), we have

$$(X^{-3})^4 = X^{-3} \times X^{-3} \times X^{-3} \times X^{-3} = \frac{1}{X^3} \times \frac{1}{X^3} \times \frac{1}{X^3} \times \frac{1}{X^3}$$

$$= \frac{1}{(X^3)^4} = \frac{1}{X^{12}} = X^{-12}.$$

So we would like to take $(-3) \times 4 = -12$, and again there was nothing special about the numbers 3 and 4 so we would substitute '*m*' for '3' and '*n*' for '4' to obtain the general result

(3.3)
$$(-m) \times n = -(m \times n),$$

for any two positive integers *m, n*

So far the rules we have obtained are very much what anybody would expect ('positive times negative is negative', 'negative times positive is negative'); however many people have a great deal of trouble with understanding why 'negative times negative is positive'. Let us again look at equation (3.1) with the object of giving meaning to, say, $(-2) \times (-3)$. We therefore consider $(X^{-2})^{-3}$. Bearing (2.4) in mind yet again, we have

$$(X^{-2})^{-3} = \frac{1}{(X^{-2})^3}; \text{ but } (X^{-2})^3 = X^{-6}$$

as we have already discovered. Thus

$$(X^{-2})^{-3} = \frac{1}{X^{-6}} = X^6, \text{ by (2.5)}.$$

It follows that, in order for (3.1) to be true, we must take $(-2) \times (-3) = 6$. Once again we observe that there was nothing special about the numbers 2 and 3, so it could be argued, in general (by replacing '2' by '*m*' and '3' by '*n*'), that

(3.4)
$$(-m) \times (-n) = m \times n,$$

for any two positive integers *m, n*.

Thus if we want to give a meaning to negative exponents, and we claim this *is* very useful, and if we want the laws of exponents (2.1), (2.2) to continue to hold, then we must assign the meaning (2.4),

$$X^{-n} = \frac{1}{X^n},$$

to a negative exponent; and we must define multiplication involving negative numbers by the rules (3.2), (3.3), and (3.4).

We claim that (3.2), (3.3), and (3.4) are reasonable on other grounds. So far as (3.2) is concerned, $m \times (-n)$ should be the same as the result of adding $(-n)$ to itself m times and this gives us the rule $m \times (-n) = -(m \times n)$. So far as (3.3) is concerned, we would not wish the value of a product to depend on the order of the factors (we take for granted that $A \times B = B \times A$, though this is not true in certain other sophisticated mathematical systems) so we would want $(-m) \times n$ to be the same as $n \times (-m)$ and this makes (3.3) a consequence of (3.2). As to (3.4), we may also establish its reasonableness this way. Suppose we want to use good mathematical notation which avoids the necessity for unnecessary thought. We would find ourselves very naturally using such an expression† (as we do in algebra) as $5 - 2X$. We would not want to have to think whether this means that we subtract 'twice X' from 5 or that we add '-2 times X' to 5. Now let $X = -7$. Then 'twice X' is -14 and subtracting -14 is, according to (1.2), the same as adding 14. So adding 14 to 5 should be the same as adding $(-2) \times (-7)$ to 5. We are led again to the rule $(-2) \times (-7) = 14$ or, more generally, (3.4).

The reader should not expect often to be confronted with situations in which it is necessary to multiply by negative numbers. This happens sometimes in graphing functions; but, except in scientific or engineering applications, such situations will be rare. All we have attempted to do is to show why the rules for adding, subtracting, and multiplying have been chosen as they have by mathematicians. Notice that these rules are really concerned with questions of 'sign' (positive or negative); the actual *algorithms* used for adding, subtracting and multiplying integers are the same algorithms used for arithmetical operations on the natural numbers, with the extra responsibility of taking account of the signs of the numbers occurring in the arithmetic and the consequent signs occurring in the answers. A good practical rule for doing a multiplication (or division) involving negative numbers is this:

first determine the sign of the answer;
then do the calculation ignoring all signs;
then attach the sign you determined.

For example, to calculate $(-13) \times 71$, the rule of signs tells us the answer is negative. $13 \times 71 = 923$, so the answer is -923.

---

†It is customary, in algebraic expressions, to suppress the product symbol '$\times$' when one, at least, of the numbers being multiplied is expressed by a letter. Thus we write $2X$ instead of $2 \times X$ (we don't write $X2$).

Thus far we have justified the rules we have adopted for multiplying negative numbers by our wish to have the laws of exponents hold for both positive and negative exponents. There are, of course, other real-life situations where these same rules seem to make sense. Consider, for example, a situation in which the temperature is falling at the rate of 2 degrees per hour and where you record the temperature at 12:00 noon to be say 37 degrees. Now three hours later you would expect the temperature to be six degrees lower than it was at noon. This change (computed by multiplying the rate of change by the time interval) can be expressed as

$$3 \times (-2) = -6,$$

where we interpret the '$-$' sign in front of the 6 to mean that the temperature has fallen. So the temperature in degrees at 3:00 p.m. would be $37 - 6$ or 31. Furthermore, you would expect that 3 hours before noon the temperature would have been six degrees warmer. In that instance the change in temperature (computed, as before, by multiplying the rate of change by the time interval) could be expressed as

$$(-3) \times (-2) = 6.$$

where the '$-$' in front of the 3 indicates that the time was in the past. Consequently we would know that the temperature in degrees at 9:00 a.m. was $37 + 6$, or 43.

So here we have another situation in which it makes sense to use the rule of signs for multiplication which we abbreviate as:

(3.5)

$$(+) \times (+) = + \qquad\qquad (+) \times (-) = -$$

$$(-) \times (-) = + . \qquad\qquad (-) \times (+) = -$$

Some people like to remember these by observing that if the signs are alike then their product is positive; if the signs are different then their product is negative.

Notice that since we now know how to multiply negative numbers, we can simplify expressions like the one that appears in Problem 16 of Exercise 5.3 more quickly. The calculation for that problem could now appear as follows (where, for pedagogical reasons we again include all of the steps that we think might be helpful to you—we expect you to abridge the calculations in your own work):

$$(A^4 \times B^{-3})^{-2} \times (A^{-2} \times B^{-3})^{-1} = A^{(4) \times (-2)} \times B^{(-3) \times (-2)}$$

$$\times A^{(-2) \times (-1)} \times B^{(-3) \times (-1)}$$

$$= A^{-8} \times B^6 \times A^2 \times B^3$$

$$= A^{-8+2} \times B^{6+3}$$

$$= A^{-6} \times B^9 \text{ or } \frac{B^9}{A^6}$$

We will make a brief mention of *division* of negative numbers in the next section.

Exercise 5.3

1. Fill in the missing values and make up a situation involving temperatures that would make sense in each case shown in the table below.

Case #	Temperature change in degrees per hour	Time change in hours	Total temperature change over the specified interval
(a)	+4	+2	+8
(b)	+4	−2	
(c)	−4	+2	
(d)	−4	−2	

Ans: (a) If the temperature is rising at a rate of 4 degrees per hour, then 2 hours from now it will be 8 degrees warmer.

2. Fill in the missing values and make up a situation involving the data given that would make sense in each case shown in the table below.

Case #	Daily price change of 1 share of IBM stock, in dollars	Time change in days	Total change of value of 1 share of IBM stock over the specified time interval
(a)	+3	+4	
(b)	+3	−4	−12
(c)	−3	+4	
(d)	−3	−4	

Ans: (b) If a share of IBM stock rises at a rate of 3 dollars each day, then 4 days ago its value was 12 dollars less than it is now.

In Problems 3 through 10 simplify the given expressions.

3. $(A^{-3} \times B^2)^{-2} \times (A^2 \times B^{-3})^2$

4. $(A^{-2} \times B^{-5})^3 \times (A^{-2} \times B^{-5})^{-3}$

5. $\dfrac{A^5 \times B^{-6}}{(A \times B^{-3})^2}$

6. $A^4 \times B^3 \times (A \times B)^{-3}$

7. $\dfrac{(A^2 \times B^{-3})^{-1}}{C^{-1}}$

8. $\left(\dfrac{A^2 \times B^{-3}}{C}\right)^{-1}$

9. $(A^{-2})^{-3}$

10. $(A^{-3})^{-2}$

11. $(B^2)^{-4}$

12. $(B^{-2})^4$

5.4 THE ARITHMETIC OF NEGATIVE FRACTIONS AND NEGATIVE DECIMALS

We begin this section by stressing again a remark we made at the end of the last section—all that is involved in adding, subtracting and multiplying integers (positive, negative or zero) that was not involved in the arithmetic of the natural numbers is the matter of determining the sign of the answer. Perhaps one should add that there may also be some question of just what arithmetic operation to perform. The examples at the end of Section 5.1 bring out this point. Thus, in Example 6, to calculate $-27 - (-82)$ we, in fact, just subtract 27 from 82. Again, to calculate $(-13) \times 6$ we compute $13 \times 6 = 78$ and attach the minus sign so that our final answer is -78.

These remarks show that, with any kind of number, if we know how to do arithmetic with positive numbers of the given kind, we can do the same arithmetic with negative numbers of the same kind, simply obeying the same laws of signs as we have developed with integers. Let us speak first of fractions; this is far more complicated

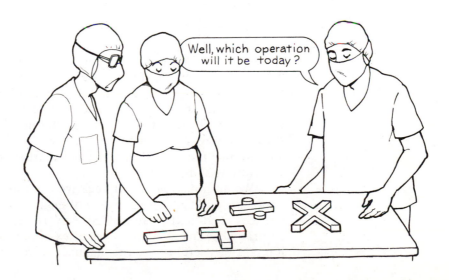

than the issue of negative decimals. Thus we would extend the validity of (1.3), (1.4) so that we have

(4.1) $X + (-Y) = X - Y,$

for any fractions X, Y; and

(4.2) $-(-X) = X,$

for any fraction X. An example of (4.1) would be

$$\frac{3}{5} + \left(-\frac{2}{5}\right) = \frac{3}{5} - \frac{2}{5} = \frac{1}{5},$$

while an example of (4.2) would be

$$-\left(-\frac{3}{4}\right) = \frac{3}{4}.$$

As to the multiplication of fractions we would have the rule of signs established in the previous section:

(4.3)
$$X \times (-Y) = (-X) \times Y = -(X \times Y),$$
$$(-X) \times (-Y) = X \times Y,$$

for any two fractions X, Y. Examples of this rule would be

$$\frac{1}{3} \times \left(-\frac{2}{5}\right) = -\frac{2}{15},$$

$$\left(-\frac{2}{7}\right) \times \frac{3}{8} = -\frac{6}{56} = -\frac{3}{28},$$

$$\left(-\frac{1}{3}\right) \times \left(-\frac{5}{9}\right) = \frac{5}{27}.$$

We hope that the reader capable of handling the arithmetic of positive fractions and understanding the rules of signs which we carefully developed in the preceding sections will have little difficulty in combining these skills in order to be able to handle the arithmetic of *all* fractions. However, that same reader may well wonder whether there is any reason to consider negative fractions at all—do they really arise in 'real life'? Do they serve as a mathematical model of anything we really wish to study?

We do not suggest that negative fractions are of supreme importance outside the domains of engineering and the physical sciences—and mathematics itself. Nevertheless, they can arise fairly naturally. Let us give two types of examples of a situation in which it might be useful to use the language of negative fractions.

Example   Measurements often involve us in simple fractions; thus $7\frac{1}{2}$ inches, $12\frac{1}{2}$ degrees, $\frac{1}{4}$ pounds. We have seen that it is natural in certain measurements (temperatures, longitudes, distances west, or south, of some fixed point) to employ negative numbers. Thus, for these measurements, it is not unreasonable also to admit the use of negative fractions—we may speak of a temperature of $-12\frac{1}{2}$ degrees Fahrenheit or we may refer to a point $6\frac{1}{2}$ miles west of Chicago as being $-6\frac{1}{2}$ miles east of Chicago.

The second type of example will be dealt with more extensively.

Example   Our rule for multiplying negative numbers was based on the laws of exponents

$$X^m \times X^n = X^{m+n},$$

$$(X^m)^n = X^{m \times n}.$$

The first law told us what meaning to give X^n when n is zero or a negative integer, and the second law told us how we should define $m \times n$ if m or n (or both) were negative. We now ask, could we attach any meaning to X^a if a were a fraciton? Let us consider first the simplest case—what should we mean by $X^{\frac{1}{2}}$? Well, if it has any meaning then, by the second law,

$$\left(X^{\frac{1}{2}}\right)^2 = X^{\frac{1}{2} \times 2} = X^1 = X.$$

So $X^{\frac{1}{2}}$ is a number whose square is X; we call such a number the *square root* of X. For example,

$$4^{\frac{1}{2}} = 2, \text{ since } \left(4^{\frac{1}{2}}\right)^2 = 4 \text{ and } 2^2 = 2 \times 2 = 4;$$

$$9^{\frac{1}{2}} = 3, \text{ since } \left(9^{\frac{1}{2}}\right)^2 = 9 \text{ and } 3^2 = 3 \times 3 = 9;$$

$$100^{\frac{1}{2}} = 10, \text{ since } \left(100^{\frac{1}{2}}\right)^2 = 100 \text{ and } 10^2 = 10 \times 10 = 100.$$

Notice that we only talk here of *positive* numbers X and the square root is itself positive. Of course, not all square roots come out 'nicely' as in our examples above, but we can always approximate to a square root as closely as we like by a decimal; for example,

$$2^{\frac{1}{2}} = 1.414\cdots, \quad 3^{\frac{1}{2}} = 1.732\cdots.$$

To see just how good these approximations are, carry out the multiplications 1.414×1.414 and 1.732×1.732 and compare the answers with 2 and 3 respectively.

We have now handled the exponent $\frac{1}{2}$; what of other fractional exponents? Let a be the positive fraction $\frac{N}{D}$. Then if X^a is to have a meaning and if the second law is to continue to work, we must have

(4.4) $$\left(X^{\frac{N}{D}}\right)^D = X^{\frac{N}{D} \times D} = X^N.$$

This means that $X^{\frac{N}{D}}$ is the Dth root of X^N; that is $X^{\frac{N}{D}}$ is the number Y such that if we multiply Y by itself D times we get X^N. Put another way, $X^{\frac{N}{D}}$ is the number Y such that is we raise Y to the Dth power we get X^N, that is $Y^D = X^N$. Let us immediately give an example to make things clearer.

$$32^{\frac{3}{5}} = 8,$$

because, if we raise 8 to the 5th power, we get 32^3; let's check this:

$$8^5 = 32768 = 32^3.$$

So we have verified that $32^{\frac{3}{5}}$ is 8, or 2^3, but the question naturally arises, how would we discover this if we didn't know it in advance? Actually we can use the second law here, but this time in the form

$$32^{\frac{3}{5}} = 32^{\frac{1}{5} \times 3} = \left(32^{\frac{1}{5}}\right)^3,$$

so we need to find $32^{\frac{1}{5}}$, that is, the fifth root† of 32. Now we know, or can easily discover, that $32 = 2 \times 2 \times 2 \times 2 \times 2$, so its fifth root is 2. Thus $32^{\frac{1}{5}} = 2$, so $32^{\frac{3}{5}} = 2^3 = 8$. The *technique* here employs the laws of exponents and the *factorization* of whole numbers (see Chapter 6).

So far we have only allowed positive fractions as exponents, but we can now remove this restriction. For we have already seen that the first law of exponents leads to the interpretation

(4.5) $$X^{-a} = \frac{1}{X^a},$$

where a is a positive integer. We simply extend (4.5) to give meaning to X^{-a} where a is a positive fraction. Thus, for example,

---

†It is customary to write $\sqrt[n]{x}$ for $x^{\frac{1}{n}}$, but to suppress the n if $n = 2$. Hence $\sqrt[5]{32} = 32^{\frac{1}{5}} = 2$ and $\sqrt{961} = 961^{\frac{1}{2}} = 31$.

$$9^{-\frac{1}{2}} = \frac{1}{9^{\frac{1}{2}}} = \frac{1}{3},$$

$$27^{-\frac{2}{3}} = \frac{1}{27^{\frac{2}{3}}} = \frac{1}{\left(27^{\frac{1}{3}}\right)^2} = \frac{1}{3^2} = \frac{1}{9}.$$

These last two examples are typical of problems that frequently appear on standardized tests. If passing that kind of test is important to you, then you may wish to study carefully the following two examples and practice on the problems given at the end of this section.

Example 1   Simplify $(32)^{\frac{4}{5}}$

Solution (i) $(32)^{\frac{4}{5}} = \left(32^{\frac{1}{5}}\right)^4 = (\sqrt[5]{32})^4 = (2)^4 = 16.$

Solution (ii) $(32)^{\frac{4}{5}} = (32^4)^{\frac{1}{5}} = (1048576)^{\frac{1}{5}}$

$$= \sqrt[5]{1048576}$$

$$= 16.$$

Solution (iii) $(32)^{\frac{4}{5}} = (2^5)^{\frac{4}{5}} = 2^{5 \times \frac{4}{5}} = 2^4 = 16.$

Example 2   Simplify $(36)^{\frac{3}{4}} \times (36)^{-\frac{1}{4}}$

Solution (i) $(36)^{\frac{3}{4}} \times (36)^{-\frac{1}{4}} = \dfrac{\left(36^{\frac{1}{4}}\right)^3}{(36)^{\frac{1}{4}}}$

$$= \frac{(2.449490)^3}{2.449490}$$

$$= (2.449490)^2$$

$$= 6.000001$$

Solution (ii) $(36)^{\frac{3}{4}} \times (36)^{-\frac{1}{4}} = \dfrac{(36^3)^{\frac{1}{4}}}{36^{\frac{1}{4}}}$

$$= \sqrt[4]{\frac{(36)^3}{36}}$$

$$= \sqrt[4]{(36)^2}$$

$$= \sqrt[4]{6 \times 6 \times 6 \times 6}$$

$$= 6.$$

Solution (iii)

$$(36)^{\frac{3}{4}} \times (36)^{-\frac{1}{4}} = (36)^{\frac{3}{4} - \frac{1}{4}}$$

$$= (36)^{\frac{1}{2}}$$

$$= \sqrt{36}$$

$$= 6.$$

Thou shalt not leave thy fraction unreduced!

Notice that solution (iii) is much simpler in each of the above examples. With a little practice you will become convinced that, if you wish to save effort, you should usually exploit the rules of exponents as far as possible, generally taking roots first and then powers (where there is a choice); and that you should only carry out messy arithmetic computations as a last resort. Observe in solution (ii) of Example 2, that it was more useful to write $(36)^2$ as $6 \times 6 \times 6 \times 6$, rather than multiplying it out and obtaining 1296. Of course you could have determined (using your calculator) that $\sqrt[4]{1296} = 6$, but proceeding as we did made the use of the calculator unnecessary.

We have pointed out to you that it is by no means always a good idea to reduce fractions (contrary to traditional preaching!). However, in problems involving fractional exponents, reducing the fraction may be very helpful. If, for example, we require $81^{\frac{10}{8}}$, then we get a mess when we take the 8th root of 81. If, however, our first step is to replace $81^{\frac{10}{8}}$ by $81^{\frac{5}{4}}$, then we use the fact that $\sqrt[4]{81} = 3$ to infer that $81^{\frac{10}{8}} = 81^{\frac{5}{4}} = 3^5 = 243$.

There is one more important point to raise before we close our discussion of negative fractions. We know that (positive) fractions arise naturally in *division*; we may, for example, regard the fraction $\frac{3}{5}$ as the result of dividing 60 by 100. Could we extend our ideas of division to include negative numbers? The answer is that we certainly can. Let us remember that division is related to multiplication by the principle that

$$X \div Y = Z \quad \text{precisely if} \quad Y \times Z = X.$$

Now we know the rule of signs for *multiplying* negative numbers and the principle above gives us the (very similar) rule of signs for dividing using negative numbers. For example, let us suppose that we wish to divide X by Y where X is negative and Y is positive. If the answer is to be Z then we must have $Y \times Z = X$. Now, by (4.3), 'positive $\times$ negative' is negative, while 'positive $\times$ positive' is positive. Since Y is positive and X is negative, this forces Z to be negative. Arguing in this way in the various cases we arrive at a division rule of signs which is exactly the same as (3.5) except that '$\times$' is replaced by '$\div$'. We may also express the rule (compare it with (4.3)) in the form

$$X \div (-Y) = (-X) \div (Y) = -(X \div Y),$$

(4.6)

$$(-X) \div (-Y) = X \div Y$$

"Blessings on thee, my child, thou art become a fraction."

Notice two important features of (4.6). First, the rule applies whether X, Y are integers or fractions. Second, the rule really allows us to regard a negative fraction as a fraction in our original sense. Originally, a fraction X had a numerator (N) and a denominator (D), and so could be regarded as the ratio of these two numbers,

$$X = \frac{N}{D}.$$

With the introduction of negative fractions, we allowed ourselves to place a minus sign to the left of a positive fraction and thus convert it to a negative fraction, thus

$$Y = -\frac{N}{D}.$$

However, with the rules (4.6) it becomes *correct* to say that

(4.7)

$$-\frac{N}{D} = \frac{-N}{D} = \frac{N}{-D}.$$

Thus, *now*, a negative fraction may again be thought of as having a numerator and a denominator. We can, for example, regard $-\frac{3}{5}$ as a fraction with numerator (-3) and denominator 5. Now everything said about dividing fractions continues to hold for negative fractions. In particular, we emphasize that the Division Algorithm for Fractions (Section 3.4) continues to hold, that is,

dividing by $\dfrac{Y}{Z}$ is the same as

(4.8)   *multiplying* by $\dfrac{Z}{Y}$,

where Y, Z are any integers.

In connection with division it is natural to ask whether, when dividing one integer, A, by another, B, it may be reasonable to present the answer in the form of a quotient and remainder (see the end of the Appendix) rather than as a fraction $\frac{A}{B}$. The answer is that it is possible and easy to do so and, in some situations, reasonable! It is customary to agree that the remainder should be positive, whatever the signs of the dividend A and the divisor B. If B is positive then, whether A is positive or negative, we can always divide A by B to get a quotient Q and a remainder R, and R will lie between 0 and $B - 1$. Thus, for example, if we divide -17 by 5 we get a quotient -4 and a remainder 3, since

$$-17 = (-4) \times 5 + 3.$$

(Of course, it would not be incorrect to talk of a quotient (-3) and a remainder (-2), since it is also true that

$$-17 = (-3) \times 5 + (-2),$$

but it would not be standard practice. It is better to have a definite rule, namely, that the remainder is never negative and is less than the divisor B, in order to get a *unique* answer to the problem of division with remainder.)

If we wish to divide by a *negative* number B, then we may again produce a quotient and non-negative remainder, but now the remainder will be less than $-B$ (remember that $-B$ is positive since B is negative). Thus, for example, if we want to divide 20 by -6, we get a quotient of (-3) and a remainder of 2, since

$$20 = (-3) \times (-6) + 2.$$

Similarly if we want to divide -37 by -4 we get a quotient of 10 and a remainder of 3 since

$$-37 = 10 \times (-4) + 3.$$

Notice that this 'quotient-remainder' form of presenting the result of a division is sensitive to whether, for example, we divide 19 by -5 or -19 by 5, although the fractions $\frac{19}{-5}$ and $\frac{-19}{5}$ represent the same (negative) number. For we could write

$$19 = (-3) \times (-5) + 4, \quad \text{quotient} = -3, \text{ remainder} = 4;$$

$$-19 = (-4) \times 5 + 1, \quad \text{quotient} = -4, \text{ remainder} = 1.$$

But this is not too surprising; after all, $\frac{3}{2}$ and $\frac{6}{4}$ are fractions representing the same number but if we divide 3 by 2 the remainder is 1, whereas if we divide 6 by 4 the remainder is 2.

To sum up, by introducing negative fractions and their arithmetic, we have extended the scope of our arithmetic with *some* practical value and with *enormous* theoretical value. The reader may not fully appreciate this theoretical value—such an appreciation would not be expected until one has gone further in mathematics—but let us at least describe the number system we have created.

The system consists of the fractions, positive, negative and zero. More precisely, it consists of the numbers represented by such fractions (since, for example, the numbers $\frac{-5}{20}$ and $\frac{2}{-8}$ are the same, even though the fractions are different). We call this the system of *rational numbers*. We may add, subtract or multiply any two rational numbers and again obtain a rational number; we may divide any rational number by any non-zero rational number and again obtain a rational number. Moreover, the system of rational numbers is the smallest system of numbers containing the natural numbers with this important property, namely that we may carry out any of the four operations of arithmetic without having to go outside the system. Indeed it is the smallest system of numbers containing the number 1 with this property!

However the system of rational numbers does not contain every number that we might want. For example, in elementary geometry we need the square root of 2 (denoted $\sqrt{2}$), since in a right triangle whose two shortest sides each have length 1 cm the hypotenuse (longest side) must be of length L cms, where $L^2 = 1^2 + 1^2$ (from the Pythagorean theorem). Thus $L^2 = 2$ and it follows that $L = \sqrt{2}$. It may be shown that there is no rational number whose square is 2 (though, as mentioned earlier, we can approximate to $\sqrt{2}$ as closely as we like by a rational number). Thus to include $\sqrt{2}$ in our number system we must go beyond the rational numbers.

Similarly, if we consider a circle then the ratio of the length of its circumference to the length of its diameter is the mysterious number π, discovered by the Greeks, and this number too, is not rational, though this is very hard to prove. Again, however, π may be approximated as closely as we like by a decimal,

$$\pi = 3.14159\cdots.$$

Any number arising in geometry may, indeed, be approximated as closely as we like by a rational number, or even by a decimal. Such 'approximatable' numbers are called *real numbers*. We may, of course, do mathematics with real numbers. For example, the number $\frac{\pi}{2}$ plays a very important role in geometry and trigonometry. However, it is a mistake to think of $\frac{\pi}{2}$ as a fraction with numerator π and denominator 2; the numerators and denominators of fractions are integers. We should think of $\frac{\pi}{2}$ as the number obtained by halving the number π or by dividing the number π by 2.

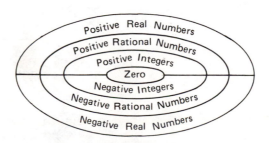

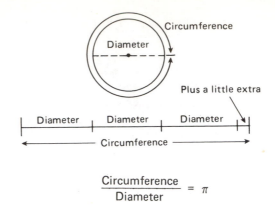

$$\frac{\text{Circumference}}{\text{Diameter}} = \pi$$

Decimals

The extension of our arithmetic to include negative decimals now presents no difficulty whatsoever. We can easily say what a negative decimal *is*. For a decimal is a fraction, so a negative decimal is a negative fraction! Thus, since $3.14 = \frac{314}{100}$, it follows that

$$-3.14 = -\frac{314}{100}.$$

If negative fractions arise in our mathematical modeling of real life then so, too, do negative decimals. In fact, they arise even more naturally, since measurements are more frequently made in decimals than in fractions. An overdraft of \$128.47 may be expressed as a balance of $-\$128.47$; a temperature of 12.83 degrees below zero Celsius may be expressed as -12.83 degrees.

How do we do arithmetic with negative decimals? The answer is again easy; we use the arithmetic of decimals we have already developed and we use the rules of signs (4.1), (4.2), (4.3), and (4.6). No really new feature enters here. The difficulty with *division,* namely that we may not be able to get an *exact* decimal answer to a division problem, remains with us when we allow negative decimals, either as dividend or divisor; it gets no worse and no better, and we may have to content ourselves with an approximate answer, to some acceptable standard of accuracy. We hope that the reader will obtain a sufficient mastery of negative decimals through the exercises.

Exercise 5.4

Simplify each of the expressions in **Problems 1 through 10.** These problems should not *require* the use of a hand calculator, but you may wish to do some of the problems several different ways in order to practice using your calculator correctly.

1. $27^{\frac{5}{3}}$

2. $64^{\frac{5}{6}}$

3. $18^{\frac{1}{2}} \times 2^{\frac{1}{2}}$

4. $45^{\frac{1}{2}} \times 5^{\frac{1}{2}}$

5. $162^{\frac{1}{4}} \times 8^{\frac{1}{4}}$

6. $32^{\frac{3}{5}} \times 64^{\frac{1}{2}}$

7. $\dfrac{32^{\frac{4}{5}} \times 16^{\frac{3}{4}}}{256^{\frac{7}{8}}}$

8. $27^{-\frac{1}{3}} \times 81^{\frac{3}{4}}$

9. $\dfrac{144^{\frac{1}{2}}}{36^{\frac{1}{2}} \times 4^{\frac{1}{2}}}$

10. $\dfrac{625^{\frac{1}{4}} \times 125^{\frac{2}{3}}}{25^{\frac{1}{2}}}$

*11.   Show that $A^{\frac{1}{2}} \times B^{\frac{1}{2}} = (A \times B)^{\frac{1}{2}}$. Generalize this to any fractional exponents.

Problems 12 through 16 will give you practice computing with negative fractions. You may notice some pattern in the answers (especially if you write the answers to problems 14 and 16 in a different, but equivalent, form).

12. $\left(\dfrac{-3}{5}\right) \times \left(\dfrac{7}{27}\right)$

13. $\left(\dfrac{-6}{25}\right) \times \left(\dfrac{-3}{4}\right)$

14. $\left(\frac{3}{5}\right) \times \left(\frac{-1}{3}\right)$

15. $\left(\frac{-26}{47}\right) \times \left(\frac{-47}{120}\right)$

16. $\left(\frac{-21}{26}\right) \times \left(\frac{2}{7}\right)$

Problems 17 through 21 will give you practice computing with negative decimals. You may wish to use your calculator to obtain the numerical part of the answer.

17. $(3395) \times (-1.6)$

18. $(-9) \times (363.5) \times (-0.5) \times (4)$

19. $(-2) \times (-43) \times (-89)$

20. $(-12.5) \times (-0.32) \times (-438.25) \times (-5)$

21. $(-32) \times (-1.5) \times (-411.5) \times (0.5)$

In each of Problems 22 through 29 express the result of the indicated division as a quotient together with a *positive* remainder. (There is little chance that you will ever be called on to execute problems of this type, certainly not on standardized tests! But if you try these problems you will gain a better understanding of the definition of division.)

22. $-22 \div 3$

23. $22 \div (-3)$

24. $-29 \div 6$

25. $29 \div (-6)$

26. $-27 \div 5$

27. $27 \div (-5)$

28. $5 \div 4$

29. $(-5) \div (-4)$

6

Factorization*

This chapter discusses the mathematics of the integers. It is more of theoretical than of practical interest and some readers may prefer to omit it.

6.1 RULES FOR DIVISIBILITY

It is often useful to know whether a given natural number m divides exactly into a given natural number n with no remainder.† For example if the wall of my living room is 42 feet long and the wallpaper I like comes in strips 8 feet wide, can I cover the wall without having to split a strip? (The answer is, of course, no.) I receive a bonus on my pay check every 3 years. I received a bonus in 1979. Will I receive a bonus in 1991? (The answer is yes, because 3 divides exactly into $12 = 1991 - 1979$.)

If m divides exactly into n, we say that m *divides* n; we also say that

m is a *factor* of n,
m is a *divisor* of n,
n is a *multiple* of m,
n is *divisible* by m,

Notice that we only use these (equivalent) terms for natural numbers m, n; in fact, in discussing questions of divisibility *we will always*

†Remember—'m divides into n' means 'n is divisible by m', because 'dividing m into n' means 'dividing n by m'.

exclude 0, so that our discussion is confined to the strictly positive integers. Thus any of the four statements above means that there is a positive integer ℓ with

(1.1) $\ell \times m = n.$

In this first section we will be concerned with rules for deciding whether n is divisible by m. Some of these rules depend on the fact that we *write* numbers using the base 10 numeral system while others depend on arithmetical properties of the numbers themselves. Let us first discuss rules of the second kind, since they will be needed to justify our rules of the first kind.

Suppose we have ascertained that 7 is a factor of 84 and a factor of 133. Can we infer that 7 is a factor of 133 + 84? of 133 − 84? If we multiply 84 by any other number, say 16, can we then infer that 7 is a factor of 16 × 84? The answer is yes to all three questions. For since 133 = 7 × 19, 84 = 7 × 12, it follows that

$$133 + 84 = 7 \times (19 + 12),$$

$$133 - 84 = 7 \times (19 - 12),$$

$$16 \times 84 = 7 \times (16 \times 12).$$

In the first two cases we use the *distributive law* for integers,

$$a \times (b + c) = (a \times b) + (a \times c);$$

in the third case we argue that $16 \times 84 = 16 \times (7 \times 12) = (16 \times 7) \times 12 = (7 \times 16) \times 12 = 7 \times (16 \times 12)$; that is, we use the *associative* and *commutative* laws for multiplication,

$$a \times (b \times c) = (a \times b) \times c \qquad \text{(associative law)},$$

$$a \times b = b \times a \qquad \text{(commutative law)}.$$

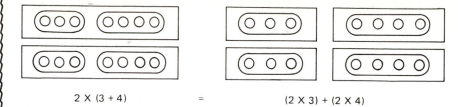

$$2 \times (3 + 4) \qquad = \qquad (2 \times 3) + (2 \times 4)$$

The general principles underlying our examples above may be stated as follows:

Arithmetic Divisibility Principles

Let m, n_1, n_2 be positive integers such that n_1 and n_2 are divisible by m, and let k be any other positive integer. Then

$$n_1 + n_2,$$

$$n_1 - n_2,$$

$$k \times n_1$$

are also divisible by m.

Our principles enable us to infer the divisibility of sums, differences and products. Can we say anything about quotients? Obviously we cannot say anything quite like the conclusion in the statement above. For example 6 is a factor of 4×3 but it is neither a factor of 4 nor a factor of 3. However, there is a principle which applies to this type of situation, but it involves the concept of a *prime number*.

We say that a positive integer p is prime if it is not equal to 1 and if its only factors are 1 and p (we will see later why we want to *exclude* 1 as a prime number). Thus the first few primes are 2, 3, 5, 7, 11, 13, $\cdots$. The prime numbers (or *primes*, as they are often called) are the multiplicative 'building blocks' of the positive integers, in the sense that *every* positive integer (except 1) is expressible as a product of primes†; thus, for example,

$$60 = 2 \times 2 \times 3 \times 5 = 2^2 \times 3 \times 5$$

$$78 = 2 \times 3 \times 13,$$

$$133 = 7 \times 19,$$

$$289 = 17 \times 17 = 17^2.$$

Moreover, such expressions are unique, provided we write the prime factors in increasing order, as we have done above, or do not bother

---

†Notice that we may still speak of a *product* of primes even if only one prime is involved. Thus $17 = 17$ expresses 17 as a product of primes. Mathematics—and mathematicians—thrive on such 'special cases' of valuable general notions.

about order. It is to achieve this uniqueness that we omit 1 from our list of prime numbers (and our list of numbers to be factorized†); obviously we could put in as many 1's as we like into a factorization if 1 were a permitted prime. The uniqueness is important because it means we can not only build all our numbers from the primes; we can also recover the prime factors from the given number—we can achieve synthesis *and* analysis.

What happens to their prime factorizations if we multiply two numbers? It should be clear that we merely multiply their prime factorizations. Thus

$$78 \times 133 = 2 \times 3 \times 13 \times 7 \times 19 = 2 \times 3 \times 7 \times 13 \times 19$$

(if we wish to write the factors in increasing order),

$$60 \times 78 = 2 \times 2 \times 3 \times 5 \times 2 \times 3 \times 13$$
$$= 2 \times 2 \times 2 \times 3 \times 3 \times 5 \times 13$$
$$= 2^3 \times 3^2 \times 5 \times 13.$$

From these examples the following *principle* should readily emerge, to go alongside our earlier principles.

If the prime p is a factor of $n_1 \times n_2$ it is a factor of at least one of n_1 and n_2.

This principle can be restated as a *divisibility principle* about quotients:

If $n = n_1 \times n_2$ and the prime p is a factor of n but not of n_1, then p is a factor of n_2, the quotient of n by n_1.

There is one final principle we wish to state before turning to those divisibility properties which depend on our use of the base 10 numeration system. Suppose b is a factor of a, and c is a factor of b, then

$$a = b \times k, \qquad b = c \times \ell,$$

so that

$$a = c \times \ell \times k,$$

which shows that *c is a factor of a.* For example since 21840 is divisible by 10 it must also be divisible by 5. We state the general result as the

Transitive Principle of Divisibility.

If $a, b, c,$ are positive integers
and if a is divisible by b
and b is divisible by c,
then a is divisible by c.

---

†Some authors use the term 'factor' instead of 'factorize'.

We turn now to divisibility properties which depend on our base 10 numeration system.

Property 1   The number n s divisible by 2 if and only if its units digit is even.

Property 2   The number n is divisible by 5 if and only if its units digit is 0 or 5.

Property 3   The number n is divisible by 10 if and only if its unit digit is 0.

Let us just show how to prove property 2. Suppose our number ends in 0, say 7840. Then n is divisible by 10 and hence by 5. Suppose our number ends in 5, say $n = 21475$. Then 21470 and 5 are divisible by 5 and so is their sum. Suppose our number ends in a digit different from 0 or 5, say $n = 378$. Now 370 is divisible by 5. If 378 were also divisible by 5, then their difference, 8, would also be divisible by 5. This obvious nonsense shows that 378 can't be divisible by 5.

To state our next properties we need some new terms. If n is any number, let n_{100} be the remainder on dividing n by 100. Thus if $n = 2376$, then $n_{100} = 76$, the number formed by the last two digits of n. Similarly define n_{1000} to be the remainder on dividing n by 1000, that is the number formed by the last three digits of n.

Property 4   The number n is divisible by 4 if and only if n_{100} is divisible by 4.

Property 5   The number n is divisible by 8 if and only if n_{1000} is divisible by 8.

The validity of Property 4 may be seen as follows. Consider the number $n = 42136$. Since 100 is divisible by 4, so is 42100. Thus 42136 is divisible by 4 if and only if 36 is divisible by 4. (Check this if you like.) We do not say (as we would with Properties 1, 2, 3) that it is *always* best to use Property 4 to test divisibility by 4. For if n is divisible by 4 it is certainly divisible by 2 (why?). Thus n is not divisible by 4 if its units digit is odd. If its units digit is even we may take half of the integer and test that number for divisibility by 2. Such a procedure combined with Property 4 may be, for some, the quickest way to test for divisibility by 4. A similar remark applies to the use of Property 5 in testing for divisibility by 8.

For our next two properties we need yet another idea. If n is a number (written as a base 10 numeral), let $s(n)$ be the *sum of its digits*. Thus if $n = 127$, $s(n) = 1 + 2 + 7 = 10$; if $n = 4285$, $s(n) = 19$.

Property 6   The integer n is divisible by 3 if and only if $s(n)$ is divisible by 3.

Property 7   The integer n is divisible by 9 if and only if $s(n)$ is divisible by 9.

We will give an idea of why Property 7 holds (the validation of Property 6 is very similar), but first let us give two examples to show its usefulness. Is 725981364 divisible by 9? Now the sum of the digits is 45. Since 45 is divisible by 9 so is our monstrous number

$$1000 = 10^3 = (2 \times 5)^3$$
$$= 2^3 \times 5^3 = 8 \times 125$$

725981364. If you didn't know that 45 is divisible by 9, you could apply the property again. The sum of the digits of 45 is 9 and surely 9 is divisible by 9(!) Is 24913842572 divisible by 9? The sum of the digits is 47 and 47 is *not* exactly divisible by 9. Neither then is our horrible number 24913842572.

Why does Property 7 work? Notice that, if we take, for example, the number 5841, then we have

$$n = 5000 + 800 + 40 + 1$$

$$s(n) = \quad 5 \;+\; 8 \;+\; 4 + 1.$$

Thus

$$n - s(n) = (5000 - 5) + (800 - 8) + (40 - 4) + (1 - 1).$$

Now

$$40 - 4 = 4 \times (10 - 1) = 4 \times 9;$$

$$800 - 8 = 8 \times (100 - 1) = 8 \times 99;$$

$$5000 - 5 = 5 \times (1000 - 1) = 5 \times 999.$$

Since 9, 99, 999 are all divisible by 9, so, by our arithmetic divisibility principles, are $40 - 4$ ($= 4 \times 9$), $800 - 8$ ($= 8 \times 99$), $5000 - 5$ ($= 5 \times 999$). So therefore is $(5000 - 5) + (800 - 8) + (40 - 4)$; but this is $n - s(n)$. Thus we conclude that

(1.2) $n - s(n)$ *is divisible by* 9.

The reader should see that this argument will work whatever number n we start with. But (1.2) implies Property 7. For, again by our divisibility principles, if $s(n)$ is divisible by 9 then, since $n - s(n)$ is divisible by 9, it follows that $n - s(n) + s(n)$, or n itself, is divisible by 9; and, if n is divisible by 9, then $n - (n - s(n))$, or $s(n)$ itself, is divisible by 9.

Property 7 goes under the attractive name of *casting out* 9's. We will see in Section 4 that 'casting out 9's' may be used in a much more far-reaching way than merely to check divisibility by 9.

Notice also that if we start with a really large number n, so that $s(n)$ is greater than 9, we may form $s(s(n))$, or even $s(s(s(n)))$, if we like to test divisibility by 9 (we did this with our number 725981364, where $s(n) = 45$, $s(s(n)) = 9$). If we go on this way repeatedly summing the digits, then the numbers divisible by 9 will be precisely those giving us 9 by this process; and the numbers divisible by 3 will be precisely those giving us 3, 6 or 9 by this process.

Exercise 6.1

1.   You are told that 598283 is divisible by 7. Which of the following numbers are divisible by 7? Give your reasons.

   (a)   598297

(b)   598213

(c)   5982834

(d)   6598283

(e)   598283598283

2. How many numbers between 0 and 1,000,000 (inclusive) are divisible by

(a)   3?

(b)   4?

(c)   5?

(d)   9?

3. Factorize as a product of prime numbers.

(a)   16384

(b)   1029

(c)   1,000,000

(d) $\dfrac{960}{32}$

(e) $(144)^2$

(f)   160 + 9

4. (a)   What is the fourth root of 3^8?

(b)   What is the cube root of 37^{78}?

5. Which of the following numbers are divisible by 3? Give your reasons.

(a)   1111   (d)   278872278

(b)   11111   (e)   123456

(c)   111111   (f)   −4176

6. Which of the following numbers are divisible by 9? Give your reasons.

(a)   428387

(b) $10^{20} - 1$

(c) $10^{20} + 3$

(d) 7^{27}

(e)   123456789

*7. X and Y are two whole numbers. X is divisible by Y and Y is divisible by X. What can you say about X and Y?

*8. Show that X is divisible by 3 if and only if X^2 is divisible by 9.

6.2 GREATEST COMMON DIVISOR (GCD)

We discussed in Section 2.3 the problem of reducing a fraction so that it is completely reduced or, as is sometimes said, in its lowest terms. Thus, for example, the fraction $\frac{24}{60}$ may be reduced to $\frac{2}{5}$ by dividing top and bottom by 12, but may not then be reduced any further. Is there a systematic way of finding this factor 12 of both 24 and 60 so that we would know how to reduce this—or any other— fraction? Notice that 12 is the biggest factor which 24 and 60 have in common, so that we are asking for the *highest common factor* or, as it is also called, the *greatest common divisor*. We will use the latter term, abbreviated to gcd.

We begin by reminding you of the notion of a prime number. This is a number p, greater than 1, whose only factors are p and 1. Thus 2, 3, 5, 7, 11, 13, 17, 19 are the first few primes. We can build up our knowledge of prime numbers by a method known as the *sieve of Eratosthenes* which we now describe. Suppose we want to discover all the primes up to 50. We write out all the numbers from 2 to 50. Now 2 is prime. We proceed to cross out all multiples of 2 except 2 itself. The smallest survivor is 3. Thus 3 is prime and we cross out all multiples of 3 except 3 itself. The smallest survivor is 5. Thus 5 is prime . . . We proceed in this way until we've finished; the survivors are the primes up to 50. Here is what the list looks like after we've crossed out the multiples of 7.

	2	3	~~4~~	5	~~6~~	7	~~8~~	~~9~~	~~10~~
11	~~12~~	13	~~14~~	~~15~~	~~16~~	17	~~18~~	19	~~20~~
21	~~22~~	23	24	~~25~~	~~26~~	~~27~~	~~28~~	29	~~30~~
31	~~32~~	~~33~~	~~34~~	~~35~~	~~36~~	37	~~38~~	~~39~~	~~40~~
41	~~42~~	43	~~44~~	~~45~~	~~46~~	47	~~48~~	~~49~~	~~50~~

And this is also, in fact, what the final list looks like! The reason why we already have the final list (up to 50) is this. Suppose a number N is not prime (we call such numbers *composite*) and suppose its smallest prime factor is p. Then $N = p \times q$, say, and q has no prime factors smaller than p (or else N would also). This means that N would have to be *at least* $p \times p$ or p^2. Thus, since 11 is the next prime after 7, the first number to get crossed out as a multiple of 11, which was not already crossed out as a multiple of a smaller prime, is 11^2 or 121 — and we only constructed our list up to 50.

Suppose then we know, or have a means of deciding, which numbers are prime. Our next step is to devise a rule for factorizing composite numbers into primes. We discussed such prime factorizations briefly in the previous section; now we want to show how the factorization can be achieved systematically. We take an example — suppose we want to factorize 840. We test the first prime 2; obviously 2 is a factor so we have

840 = 2 × 420.

We test 420 again for the factor 2; proceeding in this way we get

840 = 2 × 420

= 2 × 2 × 210

= 2 × 2 × 2 × 105.

Now 105 is not divisible by 2, so we test for divisibility by 3; we get

$$840 = 2 \times 2 \times 2 \times 3 \times 35$$

and, finally

$$840 = 2 \times 2 \times 2 \times 3 \times 5 \times 7,$$

giving us the prime factorization, which we may also write

$$840 = 2^3 \times 3 \times 5 \times 7.$$

As another example,

$$1430 = 2 \times 715$$
$$= 2 \times 5 \times 143$$
$$= 2 \times 5 \times 11 \times 13.$$

What we have described in these two examples is a *systematic* procedure—the sort of thing a machine could easily be programmed to do. We human beings rarely function best when imitating machines, so we would probably depart from the systematic procedure in any given case. We would very likely invoke the principle, indicated in the previous section, that, if we know the prime factorizations of two (or more) numbers, then we get the prime factorization of their product by putting all their prime factorizations together. Let us illustrate this with the example of 147000. Then

$$147000 = 147 \times 10 \times 10 \times 10$$
$$= 7 \times 21 \times 2 \times 5 \times 2 \times 5 \times 2 \times 5$$
$$= 7 \times 3 \times 7 \times 2 \times 5 \times 2 \times 5 \times 2 \times 5$$
$$= 2^3 \times 5^3 \times 3 \times 7^2$$
$$= 2^3 \times 3 \times 5^3 \times 7^2 \text{ (in increasing order).}$$

A calculation of this kind may also be set out as a tree, in the following way.

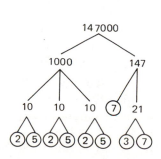

Now we approach our original problem—finding the gcd of two (or more) numbers. Again a particular example should be perfectly adequate to indicate the method. Suppose we want the gcd of 168 and 4410. We find the prime factorizations of each of them, thus

$$168 = 2^3 \times 3 \times 7,$$

$$4410 = 2 \times 3^2 \times 5 \times 7^2.$$

Let d be the gcd of 168 and 4410. Now since 2 is a factor of 168 and 4410, d as the *biggest* common factor of those numbers must itself have 2 as a factor. Could d have 2^2 as a factor? No—because 4410 doesn't have 2^2 as a factor. Proceeding in this way we find that the prime factorization of d must be *precisely* $2 \times 3 \times 7$, so that $d = 42$. We may express our result by saying 'the greatest common divisor of 168 and 4410 is 42' and we can represent the result symbolically by writing 'gcd(168, 4410) = 42'. Of course the greatest common divisor of 168 and 4410 is the same as the greatest common divisor of 4410 and 168 so it is also correct to write 'gcd(4410, 168) = 42'.

Your understanding may be increased if we explain the last example yet another way. The same result may be obtained by writing the prime factorizations of 168 and 4410 without exponents. If these factorizations are displayed, as shown, with like factors placed directly beneath their counterparts (where this is possible) then gcd(168, 4410) is obtained by multiplying together all the factors appearing in *both* factorizations (i.e., those factors connected by vertical lines).

$$168 = 2 \times 2 \times 2 \times 3 \qquad\quad \times 7$$
$$ \quad\ \ | \qquad\qquad\quad | \qquad\qquad |$$
$$4410 = 2 \qquad\qquad \times 3 \times 3 \times 5 \times 7 \times 7$$
$$ \quad\ \downarrow \qquad\qquad\quad \downarrow \qquad\qquad \downarrow$$
$$\text{gcd}(168, 4410) = 2 \qquad\qquad \times 3 \qquad\qquad \times 7$$

Again, using this method, gcd(60, 48) is easily seen to be the product of 2, 2 and 3.

$$60 = 2 \times 2 \qquad\quad \times 3 \times 5$$
$$ \quad | \quad\ | \qquad\qquad\ |$$
$$48 = 2 \times 2 \times 2 \times 2 \times 3$$
$$ \quad \downarrow \quad\ \downarrow \qquad\qquad\ \downarrow$$
$$\text{gcd}(60, 48) = 2 \times 2 \qquad\quad \times 3$$

This method, although useful for small numbers, was presented mainly to increase your understanding of the statements above that involve exponents. The method would be very tedious for extremely large numbers.

The rule should now be obvious. If we want the gcd of a and b we take the prime factorizations of a and b; if the prime p appears in *both* factorizations, we take the *smaller* power in its two appearances and that is the power of p appearing in the prime factorization of gcd(a, b). For example, if

$$a = 2^2 \times 3^2 \qquad \times 7 \times 11^2 \qquad \times 17$$

and

$$b = \qquad 3^3 \times 5 \times 7^2 \times 11^3 \times 13,$$

then

$$\gcd(a, b) = 3^2 \qquad \times 7 \times 11^2.$$

(Of course, if *no* prime appears in both factorizations, the gcd is 1.)

Once again, the method described is systematic and may well not be the best in any particular case. Let us illustrate this. It is obvious that the gcd of any number and 1 is 1, since no prime appears in the factorization of 1. So to find the gcd of 7842638 and 1, we do not have to factorize 7842638! However, there is a subtler point we can make here.

It follows from our arithmetic divisibility principles (Section 6.1) that, for example, any common factor of 60 and 48 is a factor of 60 − 48 = 12, and that any common factor of 12 and 48 is a factor of 12 + 48 = 60. More generally, the common factors of a and b are precisely the common factors of $a - b$ and b, so that

(2.1)
$$\gcd(a, b) = \gcd(a - b, b).$$

Thus if $a = b + 1$, gcd(a, b) = gcd($b + 1 - b$, b) = gcd(1, b) = 1. It would thus be absurd to factorize 71207 and 71206 to discover their gcd—it must be 1. Similarly it follows that if a, b are consecutive odd numbers then gcd(a, b) = 1, for if a, b are consecutive odd numbers then $a - b = 2$, so

(2.2)
$$\gcd(a, b) = \gcd(2, b).$$

But b is odd, so 2 is *not* a factor of b and gcd(2, b) = 1.

Furthermore, if a, b are consecutive even numbers then gcd(a, b) = 2 because if a, b are consecutive even numbers then $a - b = 2$, so equation (2.2) again holds; but now b is even so 2 *is* a factor of b and gcd(2, b) = 2.

The relationship (2.1) lies at the basis of an algorithm called the *Euclidean algorithm* for determining the gcd of two numbers. We will not go into that here but will indicate in the exercises how it works.

We close this section with a final remark. The systematic method described for finding the gcd by way of prime factorization obviously generalizes to the gcd of any collection of numbers. Although this generalization is not important for the reduction of fractions, it does find application in other places, as the exercises will show.

Exercise 6.2

1. Factorize as a product of prime numbers.
   (a)   672
   (b)   3072
   (c)   28125
   (d)   672 × 3072
   (e) 30720^3

2. Find $\gcd(x, y)$ in the following cases.
   (a) $x = 2173, y = 2173$
   (b) $x = 4346, y = 2173$
   (c) $x = 4346, y = 6519$
   (d) $x = 672, y = 3072$
   (e) $x = 672^2, y = 3072$
   (f) $x = 672^2, y = 3072^2$

3. The entry halls of the houses in S. Illyburg are to be tiled with square tiles. All the entry halls concerned have rectangular floors and their measurements are of three kinds:
   (i)   72 inches × 120 inches,
   (ii)   60 inches × 90 inches,
   (iii) 84 inches × 84 inches.
   What is the area of the largest tile that would be suitable, without any cutting, for all of these entry halls?

4. Give a rule, like that on page 225, for finding the gcd of three numbers.

*5. Suppose that, on dividing A by B, we get a quotient Q and a remainder R, so that

   $$A = B \times Q + R.$$

   Show that $\gcd(A, B) = \gcd(B, R)$. What does this mean if $R = 0$? If $R \neq 0$ divide B by R to get a quotient Q_1 and a remainder R_1. Show that $\gcd(A, B) = \gcd(R, R_1)$. Apply this procedure with $A = 5952, B = 2952$. What is R? What is R_1? What is $\gcd(A, B)$? (You should see that the process, in general, continues with remainders $R, R_1, R_2, R_3, \cdots$, until we reach a situation with a remainder R_n turning out to be 0. Then R_{n-1} is the gcd. This is the celebrated *Euclidean algorithm* for finding the gcd.)

6.3 LEAST COMMON MULTIPLE (LCM)

Let us recall the procedure for adding fractions. First, we will take an example. If we have to add $\frac{2}{9} + \frac{5}{6}$ we look for a common multiple of the two denominators 9 and 6. One such common multiple is their product, 54. Then we convert each fraction to one with denominator 54. Thus $\frac{2}{9} = \frac{12}{54}, \frac{5}{6} = \frac{45}{54}$, so that $\frac{2}{9} + \frac{5}{6} = \frac{12}{54} + \frac{45}{54} = \frac{57}{54}$. This method works in general: to add $\frac{A}{B} + \frac{C}{D}$ we find a common multiple of the two denominators and a choice of such a common denominator is the product $B \times D$. Then

$$\frac{A}{B} = \frac{A \times D}{B \times D}, \qquad \frac{C}{D} = \frac{B \times C}{B \times D},$$

so that

$$\frac{A}{B} + \frac{C}{D} = \frac{A \times D}{B \times D} + \frac{B \times C}{B \times D} = \frac{(A \times D) + (B \times C)}{B \times D}.$$

If we return for a moment to the example $\frac{2}{9} + \frac{5}{6}$, we may see that there is a smaller common multiple of 9 and 6 than their product 54, namely 18. It would plainly be more convenient to use 18 rather than 54, since this makes the arithmetic easier. Moreover, if we use an unnecessarily large common multiple the fraction we get by adding two (or more) fractions is always reducible; thus, in our example, $\frac{57}{54}$ is reducible to $\frac{19}{18}$.

This shows the advantage of finding the *least common multiple* (lcm) of the two denominators, provided there is a convenient algorithm for doing so. We now show that prime factorization provides such an algorithm.

Suppose then that we are looking for the lcm of 168 and 4410. We find their prime factorizations, thus

$$(3.1) \qquad \begin{aligned} 168 &= 2^3 \times 3 \times 7, \\ 4410 &= 2 \times 3^2 \times 5 \times 7^2. \end{aligned}$$

If a number N is to be a common multiple of 168 and 4410 it must be divisible by 2^3, since 168 is divisible by 2^3; it must be divisible by 3^2, since 4410 is divisible by 3^2; it must be divisible by 5; it must be divisible by 7^2. Thus we see that the lcm(168, 4410), is precisely the number

$$(3.2) \qquad N = 2^3 \times 3^2 \times 5 \times 7^2.$$

We denote this fact by writing lcm(168, 4410) = $2^3 \times 3^2 \times 5 \times 7^2$, or, equivalently, lcm(4410, 168) = $2^3 \times 3^2 \times 5 \times 7^2$.

The display used in Section 6.2 to compute gcd(168, 4410) may also be used to determine lcm(168, 4410). To do this, again write the factorizations of 168 and 4410 with the like factors placed directly under each other. The lcm of 168 and 4410 is then obtained by multiplying together all of the factors appearing in one, or the other, or both of the factorizations along vertical lines. Thus

$$168 = 2 \times 2 \times 2 \times 3 \qquad\qquad \times 7$$
$$4410 = 2 \qquad\qquad \times 3 \times 3 \times 5 \times 7 \times 7$$
$$\text{so lcm}(168, 4410) = 2 \times 2 \times 2 \times 3 \times 3 \times 5 \times 7 \times 7 =$$
$$= 2^3 \times 3^2 \times 5 \times 7^2.$$

Similarly lcm(60, 48) is then displayed as follows:

$$
\begin{array}{l}
60 = 2 \times 2 \qquad\qquad \times 3 \times 5 \\
\quad\quad\ \ |\ \ \ | \qquad\qquad\qquad\ | \qquad | \\
48 = 2 \times 2 \times 2 \times 2 \times 3 \\
\quad\quad\ \downarrow\ \ \downarrow\ \ \ \downarrow\ \ \ \downarrow\ \ \ \downarrow \qquad\quad \downarrow
\end{array}
$$

so lcm(60, 48) $= 2 \times 2 \times 2 \times 2 \times 3 \times 5 = 2^4 \times 3 \times 5$.

As in Section 6.2 these examples are presented so that you may get a better understanding of the statements involving exponents. It is not intended that you work these problems this way after you have mastered the exponential forms.

Now, since the lcm(168, 4410) is the product of all the factors appearing in either 168 or 4410 (or both) you should be able to see that

$$\text{lcm}(168, 4410) = 168 \times 3 \times 5 \times 7 = 168 \times 105$$

and

$$\text{lcm}(168, 4410) = 4410 \times 2^2 = 4410 \times 4.$$

This is a useful observation because it gives us two ways to compute lcm(168, 4410) and the second expression is, by far, the easier computation. A systematic procedure would have been to multiply out the expression in (3.2).

The method should now be plain. Given any two numbers a, b we find their prime factorizations. If p is a prime factor of either a or b, we take the *larger* power in its two appearances (if it is a factor of both a and b) or the power to which it appears (if it is a factor of only one of a and b), and that is the power of p appearing in the prime factorization of lcm(a, b). For example, if

$$a = 2^2 \times 3^2 \times 7 \times 11^2 \times 17$$

and

$$b = 3^3 \times 5 \times 7^2 \times 11^3 \times 13,$$

then

$$\text{lcm}(a, b) = 2^2 \times 3^3 \times 5 \times 7^2 \times 11^3 \times 13 \times 17.$$

$731 \div 7 = \ldots$ poo
$731 \div 11 = \ldots$ poo again
$731 \div 13 = \ldots$ poo
$731 \div 17 = 43$ Ho!
$731 = 17 \times 43$ Phew!
Now, $7310 \div 2 = \ldots$

This, then, is a systematic method. But, as we have so often stressed, it is often best not to use a systematic method in a particular case. It is clear, for example, that if b is itself a multiple of a, then the lcm of a and b is b itself. If it is easy to recognize that b is a multiple of a (thus $a = 731$, $b = 7310$) it would plainly be ridiculous to factorize a and b to obtain their lcm.

We also remark that the systematic method applies just as well to the lcm of three or more integers. We will give examples in the exercises so that you get practice in the technique.

Exercise 6.3

1. Find lcm(*x*, *y*) for each of the pairs of numbers (*x*, *y*) in Exercise 2 of Section 6.2. (You may, of course, leave your result as a product of factors.)

2. Carry out the following calculations.

   (a) $\dfrac{3}{20} + \dfrac{5}{48}$ (c) $\dfrac{13}{4346} + \dfrac{7}{6519}$

   (b) $\dfrac{11}{112} - \dfrac{5}{84}$ (d) $\dfrac{7}{1000} - \dfrac{1}{40} + \dfrac{11}{60}$

3. A tree trunk may be chopped up into equal pieces 16 inches long, with nothing left over. It may also be chopped up into equal pieces 28 inches long, with nothing left over. If the tree trunk is between 500 and 600 inches in length, what is its precise length? (This is an illustration of the use of the lcm. It is not a genuine application!)

4. Give a rule, like that on page 228, for finding the lcm of three numbers.

*5. Use the rules given in the text for finding gcd(*A*, *B*) and lcm(*A*, *B*), in terms of the prime factorizations of *A* and *B*, to show that

$$A \times B = \gcd(A, B) \times \text{lcm}(A, B).$$

6.4   CASTING OUT 9's

In this section we present a valuable 'check' for whole number computations. The method, known as 'casting out 9's', is extremely useful because it can be mastered quickly and in many cases it can be carried out without writing anything down. It is important to understand, however, that if the problem satisfactorily passes the casting out 9's test it does not guarantee that the computation is correct. Thus, as is often the case in non-mathematical situations, we are then in the position of knowing that something is wrong without knowing what the right answer is (this is frequently the case in situations involving politics!) This subtle, but very crucial, aspect of this check makes it important that you comprehend what the propositions and theorems in this section *mean*—even though you may not wish to master their proofs.†

  We begin by referring to our test for divisibility by 9 (Section 6.1) in which we simply add the digits of the given number. Thus, if we want to know if 2871 is divisible by 9 we add the digits of 2871, getting 18; then, since 18 is divisible by 9, so is 2871. Again, we may consider 39207; since the sum of its digits is 21, and 21 is not (exactly) divisible by 9, neither is 39207.

†We feel it would be an insult to your intelligence not to include the proofs of these results. We believe that motivated readers will be able to understand the proofs, if they study them. However, you may not feel inspired enough to expend that much effort right now. In that case, you should make up and study several numerical examples, so that you will feel comfortable with the *meaning* of each result.

$$30 \equiv 3$$

$$12 \equiv 3$$

Our explanation of this remarkable rule was the following. Suppose that $s(n)$ is the sum of the digits of the number n (so that $s(2871) = 2 + 8 + 7 + 1 = 18$, $s(39207) = 3 + 9 + 2 + 0 + 7 = 21$). Then we argued that $n - s(n)$ *is always divisible by 9.* Let us write for any two integers a, b

(4.1) $$a \equiv b$$

to mean that $(a - b)$ is divisible† by 9. Thus $40 \equiv 4$, $12 \equiv 30$, $90 \equiv 0$. What we showed in Section 1 was that

(4.2) $$n \equiv s(n).$$

We call (4.1) the relation of *congruence modulo 9.* For our purposes the following two facts will be of crucial importance. They are so important that we state them as a formal proposition.

Proposition 1   (i)   If $a \equiv b$ and $c \equiv d$, then $a + c \equiv b + d$, $a - c \equiv b - d$.

(ii)   If $a \equiv b$ and k is any integer, then $k \times a \equiv k \times b$.

Proof of (i) $(a - b)$ and $(c - d)$ are both divisible by 9. So, therefore, by our Arithmetic Divisibility Principles (Section 6.1) are their sum and difference. But

$$(a - b) + (c - d) = (a + c) - (b + d),$$
$$(a - b) - (c - d) = (a - c) - (b - d).$$

So $(a + c) - (b + d)$ is divisible by 9, which allows us to write

$$a + c \equiv b + d;$$

and $(a - c) - (b - d)$ is divisible by 9, which allows us to write

$$a - c \equiv b - d.$$

Proof of (ii) $(a - b)$ is divisible by 9. So therefore, by our Arithmetic Divisibility Principles, is $k \times (a - b)$. But

---

†Now that we have negative numbers you should be happy with the statement that -18 is divisible by 9; the quotient is the integer -2. But, if you prefer, you can always take the positive difference between a and b in deciding whether $a \equiv b$. Of course, we may write $a \equiv b$ or $b \equiv a$, to express the same relationship.

$$k \times (a - b) = (k \times a) - (k \times b).$$

So $(k \times a) - (k \times b)$ is divisible by 9, which allows us to write

$$k \times a \equiv k \times b.$$

Thus, as an example of (i), $28 \equiv 10$ and $12 \equiv 3$. We infer, correctly, that $28 + 12 \equiv 10 + 3$, or $40 \equiv 13$; and likewise that $28 - 12 \equiv 10 - 3$, or $16 \equiv 7$. As an example of (ii), $28 \equiv 10$, so we infer that $5 \times 28 \equiv 5 \times 10$, or $140 \equiv 50$.

Finally there is one more crucial fact we need; this is referred to as the *transitivity* of the congruence relation.

Proposition 2 If $a \equiv b$ and $b \equiv c$ then $a \equiv c$.

For if $(a - b)$ and $(b - c)$ are both divisible by 9, so is their sum, $(a - b) + (b - c)$, which is $(a - c)$.

We use (4.2), together with our two propositions, to establish the following result. This result is even more important, to our present purposes, so we call it a *theorem*.

Theorem 3 Let m, n be positive integers with $m \geqslant n$. Then

$$s(m + n) \equiv s(m) + s(n)$$

$$s(m - n) \equiv s(m) - s(n)$$

$$s(m \times n) \equiv s(m) \times s(n).$$

Proof We know that $s(m + n) \equiv m + n$, by (4.2). Now $m \equiv s(m)$, $n \equiv s(n)$, so that $m + n \equiv s(m) + s(n)$, by Proposition 1 (i). Our assertion that $s(m + n) \equiv s(m) + s(n)$ now follows from Proposition 2. A similar argument shows that $s(m - n) \equiv s(m) - s(n)$.

To prove the third congruence of our theorem, we make repeated use of transitivity. First $s(m \times n) \equiv m \times n$. On the other hand $m \times n \equiv m \times s(n) \equiv s(m) \times s(n)$ by two applications of Proposition 1 (ii). Thus

$$s(m \times n) \equiv s(m) \times s(n),$$

as claimed.

Examples If $m = 112, n = 23$, then $s(m) = 4, s(n) = 5, s(m + n) = s(135) = 9$, $s(m - n) = s(89) = 17, s(m \times n) = s(2576) = 20$. Then

$$9 \equiv 4 + 5.$$

$$17 \equiv 4 - 5 = -1, \text{ because 18 is divisible by 9,}$$

$$20 \equiv 4 \times 5.$$

Why is Theorem 3 so important? It is because it establishes the following practical rule. Suppose we do a complicated calculation involving addition, subtraction and multiplication, using the numbers $a_1, a_2, a_3. \cdots, a_k$, and we get the answer b. Then if we did the same calculation using the numbers $s(a_1), s(a_2), s(a_3), \cdots, s(a_k)$, we would get an answer congruent to $s(b)$. Let us give an example. You may check that

$$(328 \times (167 + 546)) - 47293 = 186571.$$

What happens if we apply s to each term of the left? Well, since

$$s(328) = 3 + 2 + 8 = 13,$$

$$s(167) = 1 + 6 + 7 = 14,$$

$$s(546) = 5 + 4 + 6 = 15,$$

$$s(47293) = 4 + 7 + 9 + 2 + 3 = 25,$$

we get the new calculation

$$(13 \times (14 + 15)) - 25$$

and this is 352. On the right hand side we see that $s(186571) = 1 + 8 + 6 + 5 + 7 + 1$, which is 28. This means all we have to do is check that $352 \equiv 28$. Of course since $352 - 28 = 324$ all we need to know is that 324 is exactly divisible by 9. Again we can apply s to 324, obtaining $s(324) = 3 + 2 + 4 = 9$.

What use is this? Let us only speak of its *practical* use (it is of great *theoretical* importance) in a situation in which you need to check the accuracy of a calculation. Suppose that you had done the original calculation and, making some small error, had produced the answer 186471 (possibly, somebody else did the calculation and reported to you that the answer was 186471). By this fairly easy check you would know the answer was wrong, because $s(186471) = 27$ and it is *not* true that $352 \equiv 27$.

But we can make our check easier still. If you have followed thus far, you should have no difficulty in believing the following. Instead of just summing the digits we can repeat the process (as described in Section 1) so that we eventually reach a number between 1 and 9. If we actually get 9, we replace it by 0 and we call the resulting number the *remainder* or *residue* modulo 9—it is, in fact, the remainder we get on dividing by 9. Thus if the residue of n is written $r(n)$, we claim that $r(742) = 4$. For $s(742) = 7 + 4 + 2 = 13$ and $s(13) = 1 + 3 = 4$. You can check that

$$742 = (9 \times 82) + 4.$$

Now we return to our problem $(328 \times (167 + 546)) - 47293$ and apply r to each term. We get the new calculation

$$(4 \times (5 + 6)) - 7$$

and this is 37. On the right hand side $r(186571) = 1$ and you may check that $37 \equiv 1$!

Even this does not exhaust the simplification we can make in carrying out this check. For we can always, at any stage replace any number n entering into our calculation by $r(n)$ and we stay within the same *congruence class* (that is, we get an answer congruent to what we would have gotten if we hadn't made the replacement). Thus, above, we could write

$$(4 \times (5 + 6)) - 7 \equiv 4 \times 2 - 7 = 1.$$

Now we have the full force of the method of 'casting out 9's'. By repeatedly 'casting out 9's' we reduce any number to its residue modulo 9, and thus may check the accuracy of an involved calculation by doing an easy one. Here's another example. You may check that

$$(624 \times (1587 - 882)) - 318176 = 121744.$$

Casting out 9's we get

$$(624 \times (1587 - 882)) - 318176 \equiv (3 \times (3 - 0)) - 8$$
$$= 9 - 8$$
$$= 1,$$

while $121744 \equiv 1$. Had somebody claimed that the result of the calculation should be 121864, you could prove the idiot wrong by observing that $121864 \equiv 4$.

What we have established, then, is an effective and usable check on a calculation involving integers. It is a check, *not* a confirmation. By this we mean that, though the check may reveal an error in the calculation it may be that an incorrect calculation escapes detection by the check of casting out 9's. Thus, for example, in our first example, the correct answer was 186571. Obviously our check could not have detected the inadvertent error of transposing some digits in the answer; for such a transposition does not affect the sum of

624 × (1587−882) −318176 = 121864

the digits and thus does not affect $s(n)$ or $r(n)$. So the incorrect answer 186751 would remain undetected by this test.

The reader may well wonder if 'casting out 9's' cannot also be used in a calculation involving division. The answer is that it can, provided that we don't divide by multiples of 3. The rule then is this. If we are dividing by a number, the residue of that number modulo 9 must be 1, 2, 4, 5, 7 or 8. We then

replace division by 1 by multiplication by 1,
replace division by 2 by multiplication by 5,
replace division by 4 by multiplication by 7,
replace division by 5 by multiplication by 2,
replace division by 7 by multiplication by 4,
replace division by 8 by multiplication by 8.

This rule is justified by observing that

$$1 \times 1 = 1 \equiv 1 \text{ modulo } 9,$$

$$2 \times 5 = 10 \equiv 1 \text{ modulo } 9,$$

$$4 \times 7 = 28 \equiv 1 \text{ modulo } 9,$$

$$5 \times 2 = 10 \equiv 1 \text{ modulo } 9,$$

$$7 \times 4 = 28 \equiv 1 \text{ modulo } 9,$$

and

$$8 \times 8 = 64 \equiv 1 \text{ modulo } 9.$$

It is thus, in principle, possible to check a calculation involving the addition, subtraction or multiplication of fractions, provided that no denominator is divisible by 3. Let us give one example. You may check that $\frac{1}{2} + \frac{2}{5} - \frac{3}{7} = \frac{33}{70}$. Casting out 9's we replace $\frac{1}{2}$ by $1 \times 5 = 5$, $\frac{2}{5}$ by $2 \times 2 = 4$, and $\frac{3}{7}$ by $3 \times 4 = 12$. Thus $\frac{1}{2} + \frac{2}{5} - \frac{3}{7}$ is replaced by $5 + 4 - 12 \equiv 5 + 4 - 3 = 6$. On the other hand $33 \equiv 6$, $70 \equiv 7$ so we replace $\frac{33}{70}$ first by $\frac{6}{7}$ and then by $6 \times 4 = 24 \equiv 6$. This checks our calculation; we would have detected the error in asserting that $\frac{1}{2} + \frac{2}{5} - \frac{3}{7} = \frac{31}{70}$.

Decimals are even easier to handle—just ignore the decimal point and carry out the check, already described, of casting out 9's; *of course this check is useless to tell us if we've put the decimal point in the right place.*

We hope that the exercises will help you to understand the method, even if you have not fully mastered the theoretical justification for it. You may wish to go back over the theory after having tried some of the exercises successfully.

Exercise 6.4

1. Find the residue modulo 9 of the following numbers without using a hand calculator.
   (a) $(2873 + 5915)^2$
   (b) $(3028 \times 473) - 4629$
   (c) $144864 \times 3475 \times 84616 \times 2378429$
   (d) 8^{92}
   (e) $7^{12} - 2^{12}$

2. Three of the following statements are false. Identify the false statements without using a hand calculator.
   (a) $7282 \times 416 = 2913832$
   (b) $4083 + (961 \times 6137) = 25184 \times 290$
   (c) $6184 + (968 \times 39) = 43936$
   (d) $512 \times 8172 \times 903 = 4001216022$

3. (a) Show that if $4 \times A \equiv 1$, then $A \equiv 7$.
   (b) Show that if $8 \times A \equiv 1$, then $A \equiv 8$.

4. Show that the following statements are false by casting out nines. Give your arguments.

   (a) $\dfrac{2}{7} + \dfrac{93}{104} = \dfrac{12}{13}$

   (b) $\dfrac{5}{19} \times \dfrac{8}{23} \div \dfrac{11}{15} = \dfrac{29}{209}$

5. (For the interested reader) Take a look at the article 'Casting out nines revisited' by Peter Hilton and Jean Pedersen, published in *Mathematics Magazine,* Vol. 54, No. 4, September 1981, and set yourself some problems on casting out 11's.

Appendix

Whole Number Algorithms

This appendix provides the reader with the requisite background skills in whole number arithmetic.

The material presented here is intended to be a *review*. It is optional reading and you may very well feel you don't need it. We suggest that you peruse these pages lightly and if you feel confident about handling these basic algorithms then pay no further attention to this appendix. However, we include this material so that if you have any feelings of insecurity about whole-number computations you can refresh your knowledge and verify that your procedures are correct. We do not go into great detail, providing extensive text material, since we feel certain you will already have developed some computational skills and that the most you will need is a reminder about how some of the algorithms are carried out. In particular, we offer you no lengthy explanations and justifications of the kind given in the main body of the text.

The first section deals with the basic number facts and includes some observations about them that may make it easier for you to remember them. The following sections pertain to the various algorithms whereby those number facts may be used to carry out addition, subtraction, multiplication and division computations. In general, the examples will appear in abbreviated and cryptic form—often in two or more apparently different formats. Where the computation is carried out and displayed in several ways you should make comparisons between each of the forms. Usually the first presentation will display the computation in full, to indicate what is really happening, and the subsequent formulations will eliminate some of the detail so that, finally, the computation is laid out in its most concise form. We assume that, as you become more familiar and comfortable with these various computations, you will want to use the shortest form, but it is sometimes useful to know the more cumbersome version if you have trouble understanding why the algorithm works or remembering how it is carried out. (Naturally, the longer format tells you more about *why* the algorithm works, and knowing why it works is what will enable you to remember it and to use it with understanding.)

Where you need to, study each example and use the formulation you are most comfortable with. It is not essential that you use the most brief version immediately. As a matter of fact you will probably use a hand calculator for most computations of this sort and in that case the *understanding* of what the computation means is far more important than the ability to perform the algorithm accurately in some specified format, and in a given allotted amount of time. In these matters speed is seldom essential, but understanding what you are doing is of vital importance, to avoid absurd answers.

A.1 BASIC NUMBER FACTS

The following two displays are traditionally known as the addition (+) and multiplication (×) tables. In each case an individual element in the body of the table represents a *number fact* relating that element to two specified elements on the boundary of that table. The first of those specified elements lies in the same horizontal row and the second lies in the same vertical column. Thus, for example, the '8' in the row labeled 3 and the column labeled 5 of the addition table represents the number fact '3 + 5 = 8'. And, likewise, the '24' in the row labeled 4 and the column labeled 6 of the multiplication table represents the number fact '4 × 6 = 24'.

+	0	1	2	3	4	5	6	7	8	9
0	0	1	2	3	4	5	6	7	8	9
1	1	2	3	4	5	6	7	8	9	10
2	2	3	4	5	6	7	8	9	10	11
3	3	4	5	6	7	8	9	10	11	12
4	4	5	6	7	8	9	10	11	12	13
5	5	6	7	8	9	10	11	12	13	14
6	6	7	8	9	10	11	12	13	14	15
7	7	8	9	10	11	12	13	14	15	16
8	8	9	10	11	12	13	14	15	16	17
9	9	10	11	12	13	14	15	16	17	18

×	0	1	2	3	4	5	6	7	8	9
0	0	0	0	0	0	0	0	0	0	0
1	0	1	2	3	4	5	6	7	8	9
2	0	2	4	6	8	10	12	14	16	18
3	0	3	6	9	12	15	18	21	24	27
4	0	4	8	12	16	20	24	28	32	36
5	0	5	10	15	20	25	30	35	40	45
6	0	6	12	18	24	30	36	42	48	54
7	0	7	14	21	28	35	42	49	56	63
8	0	8	16	24	32	40	48	56	64	72
9	0	9	18	27	36	45	54	63	72	81

You should have immediate recall of the 100 number facts displayed in each of these two tables. Fortunately, the task of remembering them is not as formidable as it might seem at first glance. By noticing certain aspects of each table you can reduce the number of items to be memorized enormously. For example, in the addition table look at the values that are symmetric to each other across the diagonal going from the upper left-hand corner to the lower right-hand corner. What do you notice about them? Of course, they occur in matching pairs, which is a manifestation of the fact that it doesn't matter which number you take first when adding two numbers (so that 2 + 3 = 3 + 2, 4 + 7 = 7 + 4, etc.). This means that if you know the value of, say, 8 + 5, you automatically know also the value of 5 + 8. This observation already reduces your task to memorizing 55 addition facts. Of those 55 facts, 10 involve the addition of 0, which is trivial, since you know that adding 0 doesn't change your original number. Then of the 45 remaining facts, those that involve '+1' and '+2' will come to you almost automatically as a result of your experience with the sequence of counting numbers. The remaining 28 facts now constitute a reasonably sized set to have to commit to memory. And even for these facts, all sorts of stratagems are available whereby your understanding of numbers can aid your memory.

A similar situation exists for the 100 facts of the multiplication table. Here, again, the values in the table occur in symmetric pairs across the diagonal running from the upper left-hand corner to the lower right-hand corner. This is because when you multiply together two numbers you get the same answer regardless of which number comes first. Thus, for example, 7 × 8 gives the same numerical answer as 8 × 7. So, as in the addition table, you need only concern yourself with the 55 facts that lie on or below the downward sloping diagonal in the table. Of those 55 facts 10 of them involve multiplication by 0 (which always yields the answer 0) and 9 of the rest involve multiplication by 1 (which doesn't change the number being multiplied, since 8 × 1 = 8, 6 × 1 = 6, etc.) Of the 36 remaining facts, 8 involve multiplying by 2—and these may easily be remembered by regarding multiplication by 2 as adding a number to itself. So we can again say that there are only 28 facts that require much effort to commit to memory.

You might have noticed that whenever you multiply a non-zero single-digit number by 9 the sum of the digits in your answer is 9. What happens when you multiply any number by an even number (that is, a number ending in 0, 2, 4, 6, or 8)? (The answer will always be an even number.) When you multiply any number by 5 what are the possibilities for the last digit (or units) in the answer? If you colored all the even numbers red in the addition table what would it look like? Why? If you colored all the even numbers red in the multiplication table what would it look like? Why?

If you have never asked yourself questions like those in the preceding paragraph you should. In fact, you should look for as many patterns as you can find within both of these tables and, in particular, for those patterns that may enable you to detect when you have made errors in your calculations. As you learn more mathematics you may wish to return to these tables and look again for patterns. It will surprise you how much more you will see as a result of your new knowledge.

A.2 ADDITION

Example 1   Compute 42 + 31.

$$
\begin{array}{r}
\text{tens} \quad \text{units} \\
\downarrow \qquad \downarrow \\
4 \qquad 2 \\
+ \quad 3 \qquad 1 \\
\hline
7 \qquad 3
\end{array}
$$

So 42 + 31 = 73.

Example 2   Compute 56 + 38.

$$
\begin{array}{r}
\text{tens} \quad \text{units} \\
\downarrow \qquad \downarrow \\
5 \qquad 6 \\
+ \quad 3 \qquad 8 \\
\hline
① \qquad 4 \quad \text{(from the units column)} \\
8 \qquad 0 \quad \text{(from the tens column)} \\
\hline
9 \qquad 4 \quad \text{(adding partial sums from the units and the tens columns)}
\end{array}
$$

Or use the shorter method, called 'carrying'. Notice where the small 1 comes from.

$$
\begin{array}{r}
5^1\ 6 \\
+\,3\ \ 8 \\
\hline
9\ \ 4
\end{array}
$$

So 56 + 38 = 94.

Example 3   Compute 412 + 157 + 86.

hundreds tens units
 ↓ ↓ ↓

$$
\begin{array}{ccc}
4 & 1 & 2 \\
1 & 5 & 7 \\
+ & 8 & 6 \\
\hline
① & 5 & \text{(from the units column)} \\
① & 4 & 0 & \text{(from the tens column)} \\
5 & 0 & 0 & \text{(from the hundreds column)} \\
\hline
6 & 5 & 5 & \text{(adding partial sums from the units,} \\
 & & & \text{tens and hundreds column)}
\end{array}
$$

Or the shorter method. Notice where the small 1's come from.

$$
\begin{array}{ccc}
4^1 & 1^1 & 2 \\
1 & 5 & 7 \\
+ & 8 & 6 \\
\hline
6 & 5 & 5
\end{array}
$$

So 412 + 157 + 86 = 655.

Example 4   Compute $439 + 467$.

Shorter

hundreds	tens	units				
↓	↓	↓		4^1	3^1	9
4	3	9		$+4$	6	7
$+4$	6	7		9	0	6

	①	6 (from the units column)
	9	0 (from the tens column)
8	0	0 (from the hundreds column)
		6 (adding units)
①	0	0 (adding tens)
8	0	0 (adding hundreds)
9	0	6 (adding partial sums from units, tens and hundreds above)

So $439 + 467 = 906$.

There are two very important laws of arithmetic relating to addition which we have used in the previous examples. We now make them explicit.

I.   If A and B are numbers than $A + B = B + A$ (the commutative law of addition).

II. If A, B and C are numbers then $A + (B + C) = (A + B) + C$ (the associative law of addition).

The first of these two laws tells us that we can add numbers in either order and the second tells us that when we wish to add three numbers we can group them in any order, so as to be always carrying out the addition of just two numbers at a time. The second law is particularly useful in some special circumstances. Suppose, for example, you spend $1.75 at one store, then $1.25 at a second store and finally $2.89 at a third. If you wish to find out your total expenditures for that shopping trip you need to compute $2.75 + $1.25 + $2.89. In this case it is decidedly simpler to compute

$$(2.75 + 1.25) + 2.89 \ = \ 4.00 + 2.89 \ = \ 6.89,$$

than to compute

$$2.75 + (1.25 + 2.89) \ = \ 2.75 + 4.14 \ = \ 6.89.$$

The answers are, of course, the same; but, as is shown here, having familiarity with the commutative and associative laws frequently saves time and effort. (Make up a similar example showing the usefulness of the commutative law as well as the associative law.) The algorithms already demonstrated, and about to be demonstrated, make repeated use of both of these basic rules. Many elementary texts (and so many elementary teachers!) refer to these rules without demonstrating their practical utility. The natural reaction of students is then to say that the rules are obvious (you're right—they are!) and to dismiss them as useless (which they aren't!).

A.3 SUBTRACTION

Subtraction is the inverse (opposite) of addition. Thus, for example, if you have 7 apples and acquire 4 more apples you have altogether 11 apples,

$$7 + 4 = 11.$$

On the other hand, if you have 11 apples and give 4 apples away you then have 7 apples,

$$11 - 4 = 7.$$

In general if A, B, C are numbers the subtraction statement $A - B = C$ corresponds to the addition statement $A = B + C$. This fact is sometimes used as a 'check'. This will be exemplified in some of the examples below. We now display the subtraction algorithms.

Example 1   Compute $467 - 325$.

		Shorter	Check
$467 =$	$400 + 60 + 7$	467	325
$- 325 =$	$- 300 - 20 - 5$	$- 325$	$+ 142$
	$100 + 40 + 2 = 142$	142	467

So $467 - 325 = 142$.

Example 2   Compute $54 - 38$.

Method 1 (stealing)

not possible to subtract here   This step is usually called 'borrowing'.†

$$\begin{array}{rcrcrcr}
 & & & \downarrow & & & \downarrow \\
54 & = & 50 + 4 & = & 40 + (10 + 4) & = & 40 + 14 \\
-38 & = & -30 - 8 & = & -30 - 8 & = & -30 - 8 \\
\hline
 & & & & & & 10 + 6 = 16.
\end{array}$$

Shorter version of Method 1 (notice that the small 4 means '40' and the small 1 goes with the 4 to mean '14'.)

$$\begin{array}{cc}
 & \textbf{Check} \\
\begin{array}{r}
\ ^4 \\
5\ \ {}^1 4 \\
-3\ \ \ 8 \\
\hline
1\ \ \ 6
\end{array}
&
\begin{array}{r}
3^1\ \ 8 \\
+1\ \ \ 6 \\
\hline
5\ \ \ 4
\end{array}
\end{array}$$

Compare the number of times you 'carry' in the addition check with the number of times you 'borrow' (or steal) in the subtraction.

Method 2 (equal additions)

$$\begin{array}{rcrcrcr}
54 & = & (50 + 4) & = & (50 + 4 + 10) & = & 50 + 4 + 10 \\
-38 & = & -(30 + 8) & = & -(30 + 8 + 10) & = & -30 - 8 - 10 \\
\hline
\end{array}$$

$$\begin{array}{rcr}
 & = & 50 + 14 \\
 & = & -40 - 8 \\
\hline
 & & 10 + 6 = 16.
\end{array}$$

Shorter version of Method 2. This method is called 'equal additions' because you add the same number to the first and second number—but you do it judiciously. In this case we added 10 to the units digit '4', making it '14', and then added 10 to the tens digit '3', making it '4'. In shorter form it appears as follows.

$$\begin{array}{r}
5\ \ ^1 4 \\
-3_1\ \ 8 \\
\hline
1\ \ \ 6
\end{array}$$ this means 4

So $54 - 38 = 16$.

---

†It would seem more reasonable to call this 'stealing' since we have no intention of returning the number 'borrowed'!

Example 3   Compute $746 - 358$.

$$
\begin{array}{cccc}
 & \text{not possible here} & \text{steal from tens} & \text{steal from hundreds} \\
 & \downarrow\quad\downarrow & \downarrow & \downarrow \\
746 = & 700 + 40 + 6 = & 700 + 30 + (10 + 6) = & 600 + (100 + 30) + (10 + 6) \\
-358 = & -300 - 50 - 8 & -300 - 50 - \quad 8 = & -300 - \qquad 50 \quad - \qquad 8
\end{array}
$$

$$
\begin{array}{rl}
= & 600 + 130 + 16 \\
= & -300 - \ 50 - \ 8 \\
\hline
 & 300 + \ 80 + \ 8 = 388.
\end{array}
$$

Short form (stealing)

$$
\begin{array}{ccc}
{}^{6} & {}^{13} & \\
\not{7} & \not{4}\,{}^{1}6 & \\
-\ 3 & 5 & 8 \\
\hline
3 & 8 & 8
\end{array}
$$

So $746 - 358 = 388$.

Short form (equal additions)

this means 6 ⟶ $\ \ 7\ {}^{1}4\ {}^{1}6$ (this means 14   this means 16)

this means 4 ⟶ $-\ 3_1\ \ 5_1\ \ 8$

$$
\begin{array}{ccc}
3 & 8 & 8
\end{array}
$$

Example 4   Compute $1003 - 27$.

$$
\begin{array}{cccc}
 & \text{there are no hundreds} & \text{steal from the} & \\
 & \text{and no tens} & \text{thousands to fill the} & \text{steal from the tens} \\
 & & \text{hundreds and tens} & \\
 & \downarrow\qquad\downarrow & & \\
1003 = & 1000 \qquad\quad + 3 = & 900 + 100 + 3 = & 900 + 90 + 13 \\
-\ \ 27 = & -20 - 7 = & -\ \ 20 - 7 = & -\ 20 - \ 7 \\
\hline
 & & & 900 + 70 + \ 6 = 976
\end{array}
$$

Short form (stealing)

$$
\begin{array}{cccc}
 & {}^{9} & {}^{9} & \\
\not{1} & \not{0}\,{}^{1}\not{0} & {}^{1}3 & \\
- & & 2 & 7 \\
\hline
9 & 7 & 6 &
\end{array}
$$

Short form (equal additions)

this method generally requires less ingenuity, since you can proceed systematically from right to left, without first worrying about the fact that sometimes there isn't anything in the next left column to borrow (or steal).

$$
\begin{array}{cccc}
1 & {}^{1}0 & {}^{1}0 & {}^{1}3 \\
-\ 1 & 1 & 2_1 & 7 \\
\hline
 & 9 & 7 & 6
\end{array}
$$

So $1003 - 27 = 976$.

A.4 MULTIPLICATION

Example 1   Compute 4 × 63.

Method 0   4 × 63 = 63 + 63 + 63 + 63 = 252

63 appears 4 times

Method 1   4 × 63 = 4 × (60 + 3) = (4 × 60) + (4 × 3)

= 240 + 12 = 252.

Method 2

```
        6   3
 ×          4
       (1)  2    (from 4 units × 3 units)
    2   4   0    (from 4 units × 6 tens)
    2   5   2    (sum of above partial products)
```

Shorter—notice where the small 1 comes from.

```
       6¹  3
 ×          4
    2   5   2
```
24 + 1

Method 3

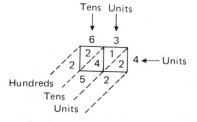

This method has the advantage that you complete all of the multiplication before you do any addition. So with this algorithm you can start on the left or right (or, indeed, in a larger problem, like Example 2, anywhere you like).

So 4 × 63 = 252.

Example 2   Compute 63 × 50

    Method 1   50 × 63 = 50 × (60 + 3)

$$= 50 \times 60 + 50 \times 3$$
$$= 3000 + 150$$
$$= 3150.$$

Method 2

```
                          6  3
                  ×       5  0
                  ─────────────
0 units × 3 units →       0  0
0 units × 6 tens  →       0  0
5 tens × 3 units  →    ①  5  0
5 tens × 3 tens   → 3  0  0  0
                  ─────────────
sum of partial    → 3  1  5  0
products
```

Short form
(Notice where the
small 1 comes from.)

```
              ¹6  3
       ×       5  0
       ─────────────
       3  1  5  0
```

Method 3

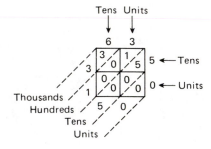

Method 4

63 × 50 = 63 × 5 × 10.

Now

    63 × 5 = 315 (use Method 3 of Example 1, say),

so

    63 × 5 × 10 = 315 × 10 = 3150.

Example 3   Compute 76 × 38.

This is now
like Example 2.

This is now
like Example 1.

Method 1

$$76 \times 38 = 76 \times (30 + 8) = (76 \times 30) + (76 \times 8)$$

$$= 2280 + 608$$

$$= 2888.$$

Method 2

```
            7  6
        ×   3  8
```

8 units × 6 units → ④ 8

8 units × 7 tens → 5 6 0

3 tens × 6 units → ① 8 0

3 tens × 7 tens → 2 1 0 0

sum of partial products → 2 8 8 8

Short form
(Notice where the 4 and 1 come from.)

①④
```
        7  6
    ×   3  8
        6  0  8
    2   2  8
    2   8  8  8
```

Method 3

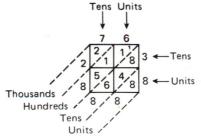

Tens Units

3 ←—Tens

8 ←—Units

Thousands
Hundreds
Tens
Units

Compare this carefully with the long form of Method 2, especially the 1 that gets carried into the hundreds position. Compare this with the short form of Method 2 to decide which you can handle more comfortably.

So 76 × 38 = 2888.

Example 4   Compute 486 × 37.

Method 1   486 × 37 = 486 × (30 + 7)

like Example 2   like Example 1

$$= (486 \times 30) + (486 \times 7)$$

$$= 14580 + 3402$$

$$= 17982$$

Method 2

			4	8	6
	×			3	7

7 units × 6 units	→				④	2
7 units × 8 tens	→			5	6	0
7 units × 4 hundreds	→		2	8	0	0
3 tens × 6 units	→			①	8	0
3 tens × 8 tens	→		2	4	0	0
3 tens × 4 hundreds	→	1	2	0	0	0
sum of partial products	1	7	9	8	2

Short form

(Notice how the small 4 and 1 relate to the computation at the left.)

$$4^6\,8^4\,6$$
$$\times\qquad 3\ 7$$
$$\overline{\qquad 3\ 4\ 0\ 2}$$
$$1\ 4\ 5\ 8$$
$$\overline{1\ 7\ 9\ 8\ 2}$$

Method 3

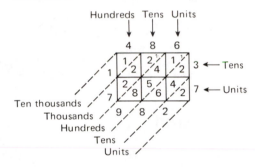

So 486 × 37 = 17982.

In the algorithms we have displayed we have made repeated use of three further laws of arithmetic, supplementing the two given in A.2. (See Section 3.5 for a fuller discussion.) The three laws are the following.

III. If A and B are numbers then $A \times B = B \times A$ (the commutative law of multiplication).

IV. If A, B and C are numbers then $A \times (B \times C) = (A \times B) \times C$ (the associative law of multiplication).

V. If A, B and C are numbers then
$A \times (B + C) = (A \times B) + (A \times C)$
(the distributive law—we say that multiplication *distributes* over addition).

Law III tells us that we can multiply two numbers in either order. Law IV says that, if we have to multiply three numbers together, we can take any two first, multiply them, and then multiply the answer by the remaining number.

Law V is a very powerful tool and often saves us a great deal of effort. Suppose, for example, that you have 6 children (in which case you will certainly need to save your energy!) and that each of them have just purchased a hot dog for $1.25 and a shake for 75¢. You could compute the total cost in either of two ways, as displayed below.

$$(6 \times 1.25) + (6 \times 0.75) = 7.50 + 4.50 = 12.00$$

or

$$(6 \times 1.25) + (6 \times 0.75) = 6 \times (1.25 + 0.75) = 6 \times 2.00 = 12.00.$$

Notice that the second, simpler computation, comes from using the *left-hand* side of the distributive law. Very often, in practice, we use the distributive law this way, making things simpler by replacing $(A \times B) + (A \times C)$ by $A \times (B + C)$. However, the multiplication algorithms actually use the law the other way round. See p. 142 for a picture justifying the truth of the distributive law.

A.5 DIVISION

Just as subtraction is the inverse (or opposite) of addition we may say that division is the inverse (or opposite) of multiplication. Thus, for example, if you know that $4 \times 3 = 12$ then you know that $12 \div 3 = 4$.

Although subtraction and division share many common features, division has several complications not encountered with subtraction. Here we will only display the algorithms for division of whole numbers by non-zero whole numbers. Furthermore we will mainly consider cases where the answer to the computation is exact—that is with no remainder. The various complications that arise, when the quotient is not exact, are discussed in detail in Chapter 3, and will only be touched on here.

Since multiplication is really repeated addition, so division is really repeated subtraction. Thus, in principle, it is possible to carry out a division by repeatedly subtracting the divisor and counting the number of times you subtract. For example, to compute $252 \div 63$ we could find the answer 4 as follows:

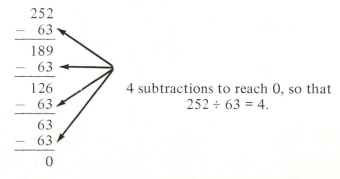

4 subtractions to reach 0, so that $252 \div 63 = 4$.

However, this method is obviously not often practical (just as it would be very impractical actually to do a multiplication like 38×26 by adding 26 to itself 38 times!). Thus we need more efficient algorithms, for division as for multiplication.

Example 1   Compute $2583 \div 7$. When the divisor is a single-digit number there is a fairly simple (short division) algorithm, which may be used instead of the standard (long division) algorithm.

Method 1 Long division.

$$
\begin{array}{r}
3\ 6\ 9 \\
7\overline{)2\ 5\ 8\ 3} \\
-2\ 1 \quad \leftarrow 3 \times 7 \\
\hline
4\ 8 \\
-4\ 2 \quad \leftarrow 6 \times 7 \\
\hline
6\ 3 \\
-6\ 3 \quad \leftarrow 9 \times 7 \\
\hline
0
\end{array}
$$

$\frac{25}{7}$ is bigger than 3, but less than 4. So the hundreds digit of the quotient is 3. Now $3 \times 7 = 21$, so subtract 21 from 25 and bring down next digit, 8, of the dividend. $\frac{48}{7}$ is bigger than 6, but less than 7. So the tens digit of the quotient is 6. Now $6 \times 7 = 42$, so subtract 42 from 48 and bring down next digit, 3, of the dividend. $\frac{63}{7}$ is exactly 9, so the units digit of the quotient is 9, and $2583 \div 7 = 369$.

Method 2 Short division.

$$
\begin{array}{r}
3\quad 6\quad 9 \\
7\overline{)2\quad 5\ {}^4 8\ {}^6 3}
\end{array}
$$

The small figure 4 results from subtracting 21 from 25. The small figure 6 results from subtracting 42 from 48. Compare the two algorithms to appreciate and understand short division.

The principle exemplified in Method 1 applies also for bigger divisors; the practice is, however, more complicated. Actually, some people are even successful in using Method 2 with large divisors. We are not recommending this to you but we will display it below when discussing Example 2.

Example 2   Compute $837 \div 31$.

Method 1 Long division.

$$
\begin{array}{r}
2\ 7 \\
31\overline{)8\ 3\ 7} \\
-6\ 2 \quad \leftarrow 2 \times 31 \\
\hline
2\ 1\ 7 \\
-2\ 1\ 7 \quad \leftarrow 7 \times 31 \\
\hline
0
\end{array}
$$

We first estimate $\frac{83}{31}$. This is about $\frac{80}{30}$ or $\frac{8}{3}$, so it is bigger than 2 and less than 3. Thus 2 appears in the tens place of the quotient. $2 \times 31 = 62$, so we subtract 62 from 83, and bring down the 7, getting 217.

We now estimate $\frac{217}{31}$. This is about $\frac{210}{30}$ or $\frac{21}{3}$ or 7. We try 7 and find it works! $7 \times 31 = 217$! (If 7 had not worked and we had 'overshot', we would next have tried 6.)

Method 2 Short division.

$$31 \overline{)\,\underset{2\,1}{837}}\overset{27}{}$$

Note that the small numbers 2, 1 can be put above or below, according to your preference. (When you get very expert you may simply carry these auxilary, numbers in your memories, in any computation.)

Method 3 Division by progressive subtraction.

$$
\begin{array}{r|l}
31\,\overline{)\,837} & \\
-620 & 20 \\
\hline
217 & \\
-155 & 5 \\
\hline
62 & \\
-62 & 2 \\
\hline
0 & 27 \leftarrow \text{Total} \\
 & \quad\;\; \text{quotient}
\end{array}
$$

By using this expanded lay-out, you have the advantage that you can 'undershoot' without doing any harm. (Overshooting is still penalized!) Thus we undershot the units in the quotient, getting first 5 and then, in the subsequent step, 2.

This method is strictly not an algorithm, but it may help you in gaining experience and confidence.

In both these examples, we have been dividing one whole number by another and getting an exact quotient. Thus we just divided 837 by 31 and obtained the quotient 27. We may write this

$$837 \div 31 = 27$$

and this means exactly the same as $837 = 31 \times 27$ (or $837 = 27 \times 31$ or $31 \times 27 = 837$ or $27 \times 31 = 837$).

What happens if we don't get an exact quotient? *The algorithms continue to work*; but now a *remainder* appears. For example, we may divide 421 by 5, getting a quotient 84 and a remainder 1. What does this mean? It means that if we multiply 84 by 5 *and then add 1,* we get 421,

$$421 = (5 \times 84) + 1.$$

Similarly if we divide 1062 by 7, we get a quotient of 151 and a remainder of 5, meaning that

$$1062 = (7 \times 151) + 5.$$

We may think of the quotients 84, 151 as *approximate* solutions of the division problem, if we like. However, the real-life meaning to be attached to these quotients depends on the context of the problem, as explained in Chapter 3.

We must especially caution you to be very careful about writing the solution to our first example as

$$421 \div 5 = 84 \text{ R } 1.$$

84 R 1 *is not a number*! The whole phrase must be read together to say 'If you divide 421 by 5 you get a quotient 84 and a remainder 1'. The *usefulness* of this statement depends, as we say, on the context (it may not be useful at all!) but the *meaning*, as a statement about numbers, is that $421 = (5 \times 84) + 1$.

Answers

Exercise 1.1, p. 31

1. (a) The '8' represents '8 dollars', or '80 dimes', or '800 cents'.

   (b) The '2' represents '2 ten-dollar bills', or '20 dollars', or '200 dimes', or '2000 cents'.

2. (a) We can conveniently carry out decimal computations in terms of dollars, dimes, and cents because the exchange between dollars and dimes (or dimes and cents) involves a factor of ten. Thus the symbol representing the former can always be changed to a symbol representing the latter by merely writing a '0' on the right of the original symbol (so 6 dollars can be exchanged for 60 dimes).

   But, changing dimes to nickels involves multiplying by 2, changing nickels to cents involves multiplying by 5 and changing dimes to cents involves multiplying by 10. The inconsistent multiplier involved in these exchanges complicates the airthmetic.

   (b) Quarters and half dollars.

3. 2 paper bills and 6 coins

4. Some of our students were able to extend the list to include $12 \times 12 \times 12$ cents. If you were able to do better than that you deserve to be congratulated!

Exercise 1.2, p. 37

1. From Table I

   (b) Accurate to the nearest millimeter

764120 millimeters
76412.0 centimeters
7641.20 decimeters
764.120 meters
76.4120 dekameters
7.64120 hectometers
.764120 kilometers

(c) Accurate to the nearest hectometer
17 hectometers
1.7 kilometers
170 dekameters
1700 meters
etc.

(d) Accurate to the nearest centimeter.
102678 centimeters
10267.8 decimeters
1026.78 meters
etc.

(e) Accurate to the nearest decimeter.
40723 decimeters
4072.3 meters
407.23 dekameters
etc.

(f) Accurate to the nearest centimeter.
91734 centimeters
9173.4 decimeters
917.34 meters
etc.

From Table II

(b) Accurate to the nearest inch.
559 inches
44 feet and 31 inches
46 feet and 7 inches
14 yards 2 feet and 31 inches
15 yards 1 foot and 7 inches
etc.

(c) Accurate to the nearest inch.
63486 inches
5289 feet and 18 inches
5290 feet and 6 inches
etc.

(d) Accurate to the nearest foot.
240 inches
20 feet
6 yards and 2 feet
etc.

(e) Accurate to the nearest inch.
63359 inches
5279 feet and 11 inches
1 inch less than a mile
etc.

(f) Accurate to the nearest inch.
63360 inches
5280 feet
1760 yards
1 mile
etc.

3. Observations will vary but they should include some concrete way of relating a centimeter, a decimeter and a meter to familiar objects. For example, 1 centimeter is about the width of most people's little finger.

5. (a) 1 minute, 1 second.

(b) The arithmetic should look like this:

$$
\begin{array}{r} 4\ 4 \\ +\ 1\ 1 \\ \hline 1\ 1\ 0 \end{array}
\qquad
\begin{array}{r} 9\ 9\ 9 \\ +\ 1\ 1\ 1 \\ \hline 1\ 1\ 1\ 0 \end{array}
$$

Case III   Case IV

(c) Suitable units are quarts, pints, cups; or dollars, half dollars, quarters; or two-dollar bills, dollars, half-dollars.

Exercise 1.3, p. 44

1. The following whole number computations enable you to infer the answers that appear below them.

$$
\begin{array}{r} 340 \\ +\ 786 \\ \hline 1126 \end{array}
$$

(a) $11.26

(e) 11.26 m

(k) 11.26 grams

$$
\begin{array}{r} 1483 \\ +\ 970 \\ \hline 2453 \end{array}
$$

(b) 24.53 m
(g) 24.53 liters
(i) 2453¢
(m) $24.53

```
  5721
+ 4279
-------
 10000
```

(c) 100.00 grams
(d) 1000.0 cm
(h) $100.00
(l) 10000 cm

```
  4167
+ 3074
-------
  7241
```

(f) 7241
(j) $72.41
(n) 724.1 decimeters

2. (a) 4310   (b) 980
    43100   8700
    431000   76000
    4310000   650000

(c) 9876   (d) 4.3
    87650   .543
    765400   .06543
    6543000 .0076543

(e) .012
    .0234
    .03456
    .045678

Exercise 1.4, p. 48

1. (a) 2.01   (d) 5.04
  (b) 3.02   (e) 6.05
  (c) 4.03   (f) 7.06

2. Any of the following:
    .3 × 26.9
    3 × 2.69
    30 × .269
    300 × .0269
    etc.

3. (a) John's plot has the larger perimeter (220 m). Mary's plot has the larger area (2400 m²).

(b) Answers will vary.
(c) The shortest perimeter will be 200 m, produced when the dimensions are 50 m by 50 m. The reason is that, of all rectangles with a given perimeter, the square encloses the largest area.

Exercise 1.5, p. 51

1. (a) 4.95
  (b) 5.95
  (c) 6.95
2. (a) 12.34
  (b) 23.45
  (c) 34.56
3. (a) 1.122
  (b) 2.233
  (c) 3.344
  (d) 4.455
4. 10% of 55.66
5. (a) $89.98
  (b) $56.66
  (c) $67.77
  (d) $78.88
6. (a) $45.55
7. (a) An item regularly selling for R dollars, on sale at a discount of $d\%$ will have a sale price of $(100 - d)\%$ of R.
  (b) Same as for Problem 5.

Exercise 2.1, p. 57

1. 2   6. 21
2. 3   7. 34
3. 5   8. 55
4. 8   9. 89
5. 13   10. 144
12. (a) Add, 2 + 3 = 5
   (b) Add, 3 + 5 = 8
   (c) 89 + 144 = 233

Exercise 2.2, p. 61

1. $6\frac{3}{8}$

2. $8\frac{4}{10}$

3. $10\frac{5}{12}$

4. $12\frac{6}{14}$

5. $14\frac{7}{16}$

6. $16\frac{8}{18}$

8. (a) $\frac{4}{10}$ or 21

   (b) $\frac{9}{20}$ of 41 $= \frac{9 \times 41}{20} = \frac{369}{20}$

   $= 18\frac{9}{20}$

9. $\frac{11}{2}$

10. $\frac{21}{5}$

11. $\frac{31}{8}$

12. $\frac{41}{11}$

13. $\frac{51}{14}$

14. $\frac{61}{17}$

16. (a) 11, 21, 31, 41, 51, 61, then 71 (add 10 each time).

   (b) 2, 5, 8, 11, 14, 17, then 20 (add 3 each time).

   (c) Problem 9: $3\frac{5}{2}$

   Problem 10: $3\frac{6}{5}$

   (d) $3\frac{11}{20} = \frac{3 \times 20 + 11}{20} = \frac{71}{20}$

17. $2\frac{52}{60}$

18. $1\frac{12}{60}$

19. $2\frac{2}{60}$

20. $3\frac{42}{60}$

21. (a) $1\frac{12}{60}, 2\frac{2}{60}, 2\frac{52}{60}, 3\frac{42}{60}$.

   A pattern is not very obvious.

   (b) $\frac{72}{60}, \frac{122}{60}, \frac{172}{60}, \frac{222}{60}$.

   The numerators increase by 50 each time.

Exercise 2.3, p. 67

1. $\frac{2}{3} = \frac{8}{12} = \frac{22}{33} = \frac{34}{51}$. $\frac{1}{2} = \frac{107}{214}$,

   $\frac{5}{7} = \frac{45}{63} = \frac{35}{49}$. $\frac{4}{9} = \frac{12}{27} = \frac{48}{108}$

2. (a) $\frac{2}{3}$

   (b) $\frac{5}{7}$

   (c) $\frac{7}{11}$

   (d) $\frac{3}{5}$

   (e) $\frac{13}{17}$

   (f) $\frac{11}{13}$

3. (a) $\frac{2}{3}, \frac{3}{5}, \frac{5}{7}, \frac{7}{11}, \frac{11}{13}, \frac{13}{17}$.

   (b) No.

   (c) 2, 3, 5, 7, 11, 13, 17, 19, 23, 29.

   (d) $\frac{17}{19}$.

4. (a) The number ends in 0, 2, 4, 6, or 8.

   (b) The sum of the digits for the numbers is 3, 3, 3, 3, 6, 6, 9, 9, 12, 18, 15. These numbers are all multiples of 3. Whenever the sum of the digits of a whole number is a multiple of 3 the number is divisible by 3.

   (c) The sum of the digits for the numbers is 9, 9, 18, 18, 27, 27. These numbers are all multiples of 9. Whenever the sum of the digits of a whole number is a multiple of 9 the number is divisible by 9.

5. (a) $\frac{87}{23}$ (c) $\frac{7}{11}$

(b) $\frac{2}{21}$ (d) $\frac{21}{143}$

Exercise 2.4, p. 73

2. (a) 1.08

(b) $\frac{12}{10} \times \frac{9}{10} = \frac{108}{100} = \frac{27}{25} = 1\frac{2}{25}$.

(c)

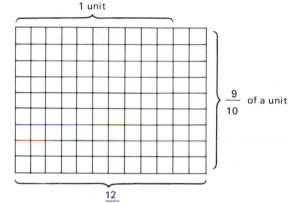

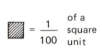

1 unit

$\frac{9}{10}$ of a unit

$\frac{12}{10}$ of a unit   108 (= 9 x 12) square units

3. (a) .123 or $\frac{123}{1000}$

(b) $\frac{12}{87}$ or $\frac{4}{29}$ or about

.1379310345

(c) $\frac{34}{65}$ or about .5230769321

(d) .234 or $\frac{234}{1000}$ or $\frac{117}{500}$

(e) .345 or $\frac{345}{1000}$ or $\frac{69}{200}$

(f) $\frac{56}{43}$ or $1\frac{13}{43}$ or about

1.302325581

(g) .456 or $\frac{456}{1000}$ or $\frac{57}{125}$.

4. (a) .123, .234, .345, .456

(b) .567

5. (a) $\frac{12}{87}, \frac{34}{65}, \frac{56}{43}$ (b) $\frac{78}{21}$.

6. (a) .1, .01, .001, .0001

(b) .1, .01, .001, .0001

7. (a) $\frac{1}{3} = .3333\cdots$,

$\frac{1}{33} = .030303\cdots$

(b) $\frac{1}{9} = .11111\cdots$,

$\frac{1}{11} = .090909\cdots$,

8. (a) $\frac{1}{3} = .3333\cdots$,

$\frac{1}{333} = .003003003\cdots$

(b) $\frac{1}{9} = .1111\cdots$,

$\frac{1}{111} = .009009009\cdots$

(c) $\frac{1}{27} = .037037\cdots$,

$$\frac{1}{37} = .027027\cdots$$

(c) $\frac{1}{41} = .0243902439\cdots,$

$$\frac{1}{2439} = .0004100041\cdots$$

9. (a) $\frac{1}{3} = .33333\cdots,$

$$\frac{1}{3333} = .00030003\cdots$$

(d) $\frac{1}{123} = .0081300813\cdots,$

(b) $\frac{1}{9} = .11111\cdots,$

$$\frac{1}{813} = .0012300123\cdots$$

$$\frac{1}{1111} = .00090009\cdots$$

(e) $\frac{1}{271} = .0036900369\cdots,$

(c) $\frac{1}{11} = .090909\cdots,$

$$\frac{1}{369} = .0027100271\cdots$$

$$\frac{1}{909} = .00110011\cdots$$

Exercise 2.5, p. 81

(d) $\frac{1}{33} = .030303\cdots,$

1. (a) $\frac{6}{9} = \frac{2}{3}$ (d) $\frac{3}{6} = \frac{1}{2}$

$$\frac{1}{303} = .00330033\cdots$$

(b) $\frac{9}{15} = \frac{3}{5}$ (e) $\frac{13}{30}$

(e) $\frac{1}{99} = .010101\cdots,$

(c) $\frac{16}{18} = \frac{8}{9}$ (f) $\frac{13}{12} = 1\frac{1}{12}$

$$\frac{1}{101} = .00990099\cdots$$

(g) $\frac{39}{30} = \frac{13}{10} = 1\frac{3}{10}$

11. The appropriate problem 10 is obtained by first writing 99999 as the product of two numbers in all possible ways (except, of course, 1×99999). Thus

$$99999 = 3 \times 33333$$
$$99999 = 9 \times 11111$$
$$99999 = 41 \times 2439$$
$$99999 = 123 \times 813$$
$$99999 = 271 \times 369$$

From each of these, in turn, we obtain

(a) $\frac{1}{3} = .3333333333\cdots,$

$$\frac{1}{33333} = .0000300003\cdots$$

(b) $\frac{1}{9} = .1111111111\cdots,$

$$\frac{1}{11111} = .0000900009\cdots$$

(h) $7\frac{6}{4} = 8\frac{2}{4} = 8\frac{1}{2}$

(i) $10\frac{12}{8} = 11\frac{4}{8} = 11\frac{1}{2}$

(j) $300\frac{7}{15}$

(k) $15\frac{22}{15} = 16\frac{7}{15}$

(l) $205\frac{263}{210} = 206\frac{53}{210}$

2. (a) 10   (c) 30
   (b) 20   (d) 40

3. (a) $15\frac{1}{2} = 15\frac{5}{10} = 15.5$

(b) $9\frac{11}{20} = 9\frac{55}{100} = 9.55$

(c) $11\dfrac{47}{50} = 11\dfrac{94}{100} = 11.94$

(d) $22\dfrac{1}{4} = 22\dfrac{25}{100} = 22.25$

(e) 16

(f) $14\dfrac{89}{100} = 14.89.$

4. (c) $\dfrac{2}{5} = \dfrac{4}{10} = .4,\ \dfrac{1}{2} = \dfrac{5}{10} = .5,$

$\dfrac{1}{25} = \dfrac{4}{100} = .04,$ so the

computation can be done as

$\begin{array}{r} 5.4 \\ 2.5 \\ + 4.04 \\ \hline 11.94 \end{array}$

(e) $\dfrac{1}{20} = \dfrac{5}{100} = .05,$

$\dfrac{1}{5} = \dfrac{2}{10} = .2,$

$\dfrac{3}{4} = \dfrac{75}{100} = .75,$ thus

$\begin{array}{r} 6.05 \\ 2.2 \\ + 7.75 \\ \hline 16.00 \end{array}$

(f) $\dfrac{3}{5} = \dfrac{6}{10} = .6,$

$\dfrac{13}{20} = \dfrac{65}{100} = .65,$

$\dfrac{16}{25} = \dfrac{64}{100} = .64,$ thus

$\begin{array}{r} 5.6 \\ 6.65 \\ + 2.64 \\ \hline 14.89 \end{array}$

5. (a) 11

(b) It is sufficient to find any set of three fractions that add to a fraction whose numerator is one less than the denominator, when the fraction is completely reduced. Then the numerator is the same as the number of horses.

Example

If you want $\dfrac{23}{24}$, then think of three numbers that add to 23, say 12, 8 and 3. Then you know that $\dfrac{12}{24} + \dfrac{8}{24} + \dfrac{3}{24} = \dfrac{23}{24}$. So you then cleverly disguise these fractions by reducing them to $\dfrac{1}{2}, \dfrac{1}{3}$ amd $\dfrac{1}{8}$.

Note that you could have used any other three numbers that would sum to 23. For example, since $10 + 9 + 4 = 23$, you could have used $\dfrac{10}{24}, \dfrac{9}{24}, \dfrac{4}{24}$ or their equivalent reduced forms of $\dfrac{5}{12}, \dfrac{3}{8}$ and $\dfrac{1}{6}$.

Either of these sets of fractions will work with 23 horses.

6. (a) $2\dfrac{2}{4} = 2\dfrac{1}{2}$

(b) $6\dfrac{1}{12}$

(c) $8\dfrac{3}{40}$

(d) $6 + \left(\dfrac{1}{4} - \dfrac{3}{8}\right)$

$= 6 + \left(\dfrac{2}{8} - \dfrac{3}{8}\right)$

$= 5 + \left(\dfrac{10}{8} - \dfrac{3}{8}\right)$

(by borrowing)

$= 5\dfrac{7}{8}$

(e) $9\dfrac{19}{21}$

(f) $69 \frac{8}{10} = 69 \frac{4}{5}$ (or, if you noticed the decimal equivalents, $142.1 - 72.3 = 69.8$, using decimal arithmetic)

7. $8\frac{3}{4}$ inches

Exercise 2.6, p. 89

1. $\frac{1}{3} \times \frac{1}{2} = \frac{1}{6}$. The experiments may appear, at first glance, to be different. If you throw a single die twice the list of possible outcomes is exactly the same as if you had thrown two dice once. This idea is useful to remember in case you are asked to throw, say, 3 dice.

2. (a) $\frac{1}{6} \times \frac{1}{6} = \frac{1}{36}$

    (b) $\frac{1}{36}$

    (c) $\frac{1}{36}$

    (d) $\frac{1}{36}$

    (e) $\frac{1}{36}$

    (f) The sum of the answers to parts (a) through (e), or $\frac{5}{36}$

3. Yes. Think of playing a game in which you win if you roll a die twice and get two consecutive numbers. When you roll the first die, if either a 1, 2, 3, 4, or 5

shows you are still 'in the running' so to speak. (If you roll a 6 you might as well quit because there is no way to get a 7 on the second roll.) Thus the probability of a successful first roll of the die is $\frac{5}{6}$. Now, if you are still in the running there is only one possible choice for the second roll. So you then have a probability of $\frac{1}{6}$ of succeeding on the second roll of the die.

4. (a) $\frac{1}{6}$

    (b) $\frac{1}{36}$

5. $\frac{15}{36} = \frac{5}{12}$

6. The first term, $\frac{1}{6} \times \frac{5}{6}$, is the probability that you will roll a 1 and then that you will roll a number bigger than a 1. Likewise, the second number is the probability that you will roll first a 2 and then a number bigger than a 2, etc. But, since you can succeed by rolling either a 1 or a 2 or a 3 or a 4 or a 5 on the first roll, you then add the probabilities of those distinct outcomes.

7. $\frac{7}{12}$

8. 1 − (probability an event doesn't happen) = (probability an event does happen).

    So the student should receive credit for the answer.

Exercise 2.7, p. 103

1-10   For problems 1 through 10.

Problem number	I Number of favorable outcomes	II Number of un-favorable outcomes	III Total number of outcomes	IV Probability of a favorable outcome	V Probability of an unfavorable outcome	VI Odds on a favorable outcome	VII Odds on an unfavorable outcome
1.	2	4	6	$\dfrac{2}{6}$	$\dfrac{4}{6}$	$\dfrac{2}{4}$	$\dfrac{4}{2}$
2.	2	5	7	$\dfrac{2}{7}$	$\dfrac{5}{7}$	$\dfrac{2}{5}$	$\dfrac{5}{2}$
3.	7	3	10	$\dfrac{7}{10}$	$\dfrac{3}{10}$	$\dfrac{7}{3}$	$\dfrac{3}{7}$
4.	12	40	52	$\dfrac{12}{52}$	$\dfrac{40}{52}$	$\dfrac{12}{40}$	$\dfrac{40}{12}$
5.	14	22	36	$\dfrac{14}{36}$	$\dfrac{22}{36}$	$\dfrac{14}{22}$	$\dfrac{22}{14}$
6.	9	18	27	$\dfrac{9}{27}$	$\dfrac{18}{27}$	$\dfrac{9}{18}$	$\dfrac{18}{9}$
7.	60	340	400	$\dfrac{60}{400}$	$\dfrac{340}{400}$	$\dfrac{60}{340}$	$\dfrac{340}{60}$
8.	13 or 312	2 or 48	15 or 360	$\dfrac{13}{15}$ or $\dfrac{312}{360}$	$\dfrac{2}{15}$ or $\dfrac{48}{360}$	$\dfrac{13}{2}$ or $\dfrac{312}{48}$	$\dfrac{2}{13}$ or $\dfrac{48}{312}$
9.	10	16	26	$\dfrac{10}{26}$	$\dfrac{16}{26}$	$\dfrac{10}{16}$	$\dfrac{16}{10}$
10.	N	$D-N$	D	$\dfrac{N}{D}$	$\dfrac{D-N}{D}$	$\dfrac{N}{D-N}$	$\dfrac{D-N}{N}$

11. (a) They equal the number in column III.
(b) Their sum is 1.
(c) Their product is 1.

12.

				III			
A.	3	7	10	$\frac{3}{10}$	$\frac{7}{10}$	$\frac{3}{7}$	$\frac{7}{3}$
B.	1	7	8	$\frac{1}{8}$	$\frac{7}{8}$	$\frac{1}{7}$	$\frac{7}{1}$
C.	2	6	8	$\frac{2}{8}$	$\frac{6}{8}$	$\frac{2}{6}$	$\frac{6}{2}$
D.	3	6	9	$\frac{3}{9}$	$\frac{6}{9}$	$\frac{3}{6}$	$\frac{6}{3}$
E.	2	7	9	$\frac{2}{9}$	$\frac{7}{9}$	$\frac{2}{7}$	$\frac{7}{2}$
F.	3	5	8	$\frac{3}{8}$	$\frac{5}{8}$	$\frac{3}{5}$	$\frac{5}{3}$

(In A, B, E, F you can't be certain of the first 3 numbers, only of their ratios. We've given the smallest numbers which work.)

13. (f) A loss of 0.11 dollars
(g) $\frac{2}{9} \times 1000 \times \$10 = \$2,222.22$
(h) $\frac{7}{9} \times 1000 \times \$3 = \$2,333.33$
(i) $\$2,333.33 - \$2,222.22$
$= \$111.11$
(j) 2 to 7 or $\frac{2}{7}$. You should pay $2.86 (to the nearest cent).

14. (a) The probability you win is $\frac{18}{38}$, the probability you lose is $\frac{20}{38}$, thus the true odds should be 18 to 20.
(b) Because the official odds are 1 to 1 the result when you win is $1.00 (or +1) and the result when you lose is −$1.00 (or −1).
Now use $R_1 = 1, R_2 = -1$ and (from part (a)) $P_1 = \frac{18}{38}, P_2 = \frac{20}{38}$ to evaluate the expectation E as follows:
$$E = (1) \times \frac{18}{38} + (-1) \times \frac{20}{38}$$
$$= -\frac{2}{38}.$$

Thus each time you stake one dollar you will lose, on the average, $\frac{2}{38}$ of a dollar (about 5.3¢).

15. (a) 15, 7, 2
(b) Approximately 21¢

16. $\frac{81}{92}$

Exercise 2.8, p. 109

1.

	A	B	Comparison
Game I	$\frac{1}{3}$	$\frac{2}{6}$	equal
Game II	$\frac{2}{3}$	$\frac{2}{3}$	equal
Total	$\frac{3}{6}$	$\frac{4}{9}$	A's average is better than B's average since $\frac{3}{6} = \frac{9}{18}$ and $\frac{4}{9} = \frac{8}{18}$.

2. (a) $A = \frac{40}{90}$, $B = \frac{30}{60}$, $C = \frac{50}{90}$, $D = \frac{70}{150}$

(b) A, B and C are all equal to each other and D's average is better than each of the others. Or, $A = B = C < D$, since $A = B = C = \frac{1}{3} = \frac{11}{33}$, and $D = \frac{4}{11} = \frac{12}{33}$.

(c) A, B and C are all equal to each other and D's average is better than each of the others. Or, $A = B = C < D$, since $A = B = C = \frac{2}{3} = \frac{8}{12}$. and $D = \frac{3}{4} = \frac{9}{12}$.

(d) Since $\frac{4}{9} = \frac{40}{90}$ (A's average)
$\frac{1}{2} = \frac{45}{90}$ (B's average)
$\frac{5}{9} = \frac{50}{90}$ (C's average)
$\frac{7}{15} = \frac{42}{90}$ (D's average)

we see that C's average is the highest, followed by B, D and A in that order.

3. (a) If X is the hit advantage over the first half of the season and Y is the hit advantage over the second half of the season, then $X + Y$ is the hit advantage over the whole season. Thus if player A has a bigger hit advantage than player B over the first half of the season and over the second half of the season player A has a bigger hit advantage than player B over the whole season.

(b) It could be argued that, between two good players, the player with more opportunities to bat would be likely to have the bigger hit advantage.

(c) No.

4. Hint: Use $B \times D$ as a common denominator.

5. (a) The time going (in hours) is $\frac{20}{10} = 2$. The time returning (in hours) is $\frac{20}{5} = 4$. Total time (in hours) = 6.

$$\frac{\text{distance traveled (in miles)}}{\text{time used for trip (in hours)}}$$

$$= \frac{40}{6} = \frac{20}{3} = 6\frac{2}{3}.$$

So the answer is $6\frac{2}{3}$ mph.

(b) The time going (in hours) is $\frac{40}{10} = 4$. The time returning (in hours) is $\frac{40}{5} = 8$. Total time (in hours) = 12.

$$\frac{\text{distance traveled (in miles)}}{\text{time used for trip (in hours)}}$$

$$= \frac{80}{12} = \frac{20}{3} = 6\frac{2}{3}.$$

So, again, we obtain the answer of $6\frac{2}{3}$.

(c) $\frac{2 \times 10 \times 5}{10 + 5} = \frac{100}{15} = \frac{20}{3}$

$$= 6\frac{2}{3}.$$

So we infer the average speed for the double journey is $6\frac{2}{3}$ mph.

(d) 7 mph. If the speed from A to B equals the speed from B to A, then this is also the average speed for the double journey.

(e) The average speed is greater if you jog in both directions at 7 mph. The average of the speeds is greater if you jog one way at 10 mph, and the other at 5 mph ($7\frac{1}{2}$ against 7).

6. $A \bigcirc B = \frac{2 \times A \times B}{A + B}$.

Thus, $\frac{1}{A \bigcirc B} = \frac{A + B}{2 \times A \times B}$

$$= \frac{1}{2}\left(\frac{A}{A \times B} + \frac{B}{A \times B}\right)$$

$$= \frac{1}{2}\left(\frac{1}{B} + \frac{1}{A}\right).$$

Exercise 3.1, p. 114

1. Will vary, but some possibilities are: $48 \div 6 = \square$

48/6

6)48 ,

etc.

2. Some typical examples:

(a) A mother has 10 marbles and wishes to give an equal number to each of her three children. How many does each receive?

(b) An employer has 27 pages of material to be typed. If she has 4 equally efficient

typists, how many pages
should she have each one
type?

(c) See Problem 5 of Section
2.5.

3. The answers given here are for
the problems given above. Your
answers will, of course, be different, since they correspond to
your problems.

(a) Mother may decide to give
each child three marbles and
put one away (as insurance
for when a child loses a
marble).

(b) Give 6 pages to all four
typists and assign the other
three pages, one at a time, to
the typists who finish their
work first.

(c) See the solution to Problem
5, Section 2.5.

Exercise 3.2, p. 121

1. (a) 2.8
    (b) $42 = (15 \times 2) + 12$.

2. (a) 4.625
    (b) $37 = (8 \times 4) + 5$.

3. (a) 15.73333333
    (b) $472 = (30 \times 15) + 22$.

4. (a) 26.66666667
    (b) $560 = (21 \times 26) + 14$.

5. (a) 15.5375
    (b) $4972 = (320 \times 15) + 172$.

6. (a) 16.98920086
    (b) $7866 = (463 \times 16) + 458$.

7. (a) 3.451957295
    (b) $970 = (281 \times 3) + 127$.

8. (a) 41.04350161
    (b) $76423 = (1862 \times 41) + 81$.

9. (a) 3.996402878
    (b) $4444 = (1112 \times 3) + 1108$.

10. (a) 99.01980198
    (b) $10001 = (101 \times 99) + 2$

11. (a) 94
    (b) $6392 = (68 \times 94) + 0$

12. (a) 324.1491086
    (b) $400000 = (1234 \times 324) + 184$

13. Since $42 = (15 \times 2) + 12$ we see
that all fifteen patrons will have
at least 2 guests and 12 patrons
will have one additional guest. So
the answer is 12.

14. 4

15. 16

16. She cannot follow the instructions. She might, tactfully, point
out to her supervisor that since
$560 = (21 \times 26) + 14$, the
closest she can come to following the instructions is to first
send 26 new employees to all
21 departments and then send
an extra employee to 14 of the
departments.

17. Since $4972 = (320 \times 15) + 172$
you can buy 15 shares and have
$172 left over.

18. (a) Each employee receives
$16.99 in June.
    (b) July's account will be
debited 37¢.

19. Since $127 = (42 \times 3) + 1$ it will
be possible to cut 42 table
cloths from the 127 feet. Hence
the total number of table cloths
will be 281 + 42, or 323. (Of
course, this problem could easily
have been done with no reference
to Problem 7(b).)

20. $42.00

21. $3\frac{1108}{1112}$ or $3\frac{277}{278}$

22. She can purchase 198 bottles
and have 2¢ left over.

23. 94 feet

24. 325, and the last one will consist
of only 184 seats.

Exercise 3.3, p. 129

1. (a) 300   (b) 300
2. (a) 30   (b) 30
3. (a) 500   (b) 500
4. (a) 70   (b) 70
5. (a) 1500   (b) 1500
6. (a) $(100 \times A) \div B$
   (b) $(10 \times A) \div B$
7. (a) True   (b) True
   (c) False, since, for example,
   $(10 - 3) + 4 = 7 + 4 = 11$,
   while $10 - (3 + 4) =$
   $10 - 7 = 3$.
   (d) True   (e) True
8. (a) 3   (d) 12
   (b) 6   (e) 15
   (c) 9
9. Any division problem with an answer of 18
10. (b) Approximately $46 \div 2$, or 23.
    Ans: 27.06
    (c) Approximately $60 \div 20$, or 3.
    Ans: 2.65
    (d) Approximately $900 \div 30$, or 30.
    Ans: 24.32
    (e) Approximately $75 \div 25$ (since this is the same as $.75 - .25$ from Problem 7(d), or 3.
    Ans: 3.17

Exercise 3.4, p. 136

1. (a) $\frac{5}{6}$ (d) $\frac{44}{45}$
   (b) $\frac{14}{15}$ (e) $\frac{65}{66}$
   (c) $\frac{27}{28}$ (f) $\frac{90}{91}$

2. (a) Each successive answer differs from 1 by a smaller amount than its predecessor. Observe, too, that in each case the denominator equals the numerator plus one.
   (b) $\frac{7}{15} \div \frac{8}{17} = \frac{7}{15} \times \frac{17}{8} = \frac{119}{120}$,

   $\frac{8}{17} \div \frac{9}{19} = \frac{8}{17} \times \frac{19}{9} = \frac{152}{153}$,

   $\frac{9}{19} \div \frac{10}{21} = \frac{9}{19} \times \frac{21}{10} = \frac{189}{190}$.

   The observation still holds true.

3. (a) $\frac{20}{3}$ or $6\frac{2}{3}$

   (b) $\frac{3}{20}$

   (c) $\frac{18}{21}$ or $\frac{6}{7}$

   (d) $\frac{21}{18}$ or $\frac{7}{6}$ or $1\frac{1}{6}$

   (e) $\frac{3}{2}$

   (f) $\frac{6}{3}$ or 2

4. For all whole numbers a, b,

   $$a\frac{1}{b} \div b\frac{1}{a} = \frac{a}{b}$$

5. (a) $56.1 \div 10 = \frac{561}{10} \div \frac{10}{1}$

   $= \frac{561}{10} \times \frac{1}{10}$

   $= \frac{561}{100}$

   $= 5\frac{61}{100}$ or 5.61

   (b) $\frac{327}{10} \div \frac{100}{1} = \frac{327}{10} \times \frac{1}{100}$

   $= \frac{327}{1000} = .327$

(c) $\dfrac{43}{1} \div \dfrac{10}{1} = \dfrac{43}{1} \times \dfrac{1}{10} = \dfrac{43}{10}$

$= 4\dfrac{3}{10} = 4.3$

(d) $\dfrac{2132}{10} \div \dfrac{100}{1} = \dfrac{2132}{10} \times \dfrac{1}{100}$

$= \dfrac{2132}{1000} = 2.132$

6. (a) $\dfrac{12}{20}$ or $\dfrac{3}{5}$

(b) $\dfrac{24}{30}$ or $\dfrac{4}{5}$

(c) $\dfrac{40}{42}$ or $\dfrac{20}{21}$

(d) $\dfrac{60}{56}$ or $\dfrac{15}{14}$ or $1\dfrac{1}{14}$

(e) $\dfrac{84}{72}$ or $\dfrac{7}{6}$ or $1\dfrac{1}{6}$

(f) $\dfrac{112}{90}$ or $\dfrac{56}{45}$ or $1\dfrac{11}{45}$

7. $\dfrac{9}{11} \div \dfrac{10}{16} = \dfrac{9}{11} \times \dfrac{16}{10} = \dfrac{144}{110} = \dfrac{72}{55}$

or about 1.31.

$\dfrac{10}{12} \div \dfrac{11}{18} = \dfrac{10}{12} \times \dfrac{18}{11} = \dfrac{180}{132} = \dfrac{15}{11}$

or about 1.36.

$\dfrac{11}{13} \div \dfrac{12}{20} = \dfrac{11}{13} \times \dfrac{20}{12} = \dfrac{220}{156} = \dfrac{55}{39}$

or about 1.41.

Yes, each answer is still larger than its predecessor, but none of the answers in this sequence will ever reach the number 2, although they get very, very close. You may check as many examples as you wish.

8. $17\dfrac{1}{2} \div \dfrac{3}{4} = \dfrac{35}{2} \times \dfrac{4}{3} = \dfrac{70}{3} = 23\dfrac{1}{3}$,

so 23 costumes can be made.

9. We could calculate $23 \times \dfrac{3}{4}$, getting $17\dfrac{1}{4}$, and conclude $\dfrac{1}{4}$ yard is left over. Alternatively, we could say that the 'remainder' of $\dfrac{1}{3}$ is '$\dfrac{1}{3}$ of a costume length', or $\dfrac{1}{3}$ of $\dfrac{3}{4}$ yard, or $\dfrac{1}{4}$ yard.

Exercise 4.1, p. 150

1. You could afford lists (b) or (c).

2. (a) REDWOOD QUARTERLY, MUSICAL REVIEWS & SEE
   or
   REDWOOD QUARTERLY, MUSICAL REVIEWS & TALK

   (b) WASHINGTON PILLAR

3. (a) Yes

   (b) Yes

   (c) No, you could go about 300 miles before 10:00 A.M. (if you had filled up your tank the previous day), about 200 miles between 10:00 A.M. and 2:00 P.M., and then your gas supply would only allow about 300 miles more.

4. Yes, it is cheaper to fly, in this case.

5. (a) Brand Y is the better buy.

   (b) The 200 sq. ft package is the better buy.

   (c) The 22 oz bottle is a much better buy.

   (d) The 12 oz is the better buy, and the 6 oz is certainly much more expensive than either of the larger sizes.

   (e) The single 8 oz package is a much better buy.

   (f) Buying $2\dfrac{1}{2}$ dozen eggs results in a very small saving. If you use fewer than a dozen eggs per week it's probably better to buy the smaller quantity.

6. (a) The 22 oz Brand A
   (b) The 32 oz Brand B
   (c) The 32 oz Brand B
   (d) Answers may vary. They should include effectiveness of the detergent, convenience of size of container, convenience of shape of container, policy of manufacturing company on issues of importance to you.

Exercise 4.2, p. 157

1. 40, 42, 42.1, impossible to round to the nearest hundredth.
2. 190, 189, 188.8, 188.76
3. 200, 199, 198.8, .198.76
4. 10, 8, 8.4, 8.43
5. 0, 0, .0, .01 or .02
6. 98000, 98000, impossible to round to the nearest tenth or nearest hundredth
7. All four required roundings are impossible.
8. 10, 17, 17.3
9. 350, 358, 358.2
10. 180, 182, 182.1
11. 180, 180, 180.0
12. 30, 38, 38.4
13. 20, 18, 17.4
14. 360, 359, 358.3
15. 190, 183, 182.2

16. 190, 181, 180.1
17. 40, 39, 38.5
18. (a) Yes, 58 to the nearest 10 is 60 in both cases.
    (b) Yes, 52 to the nearest 10 is 50 in both cases.
    (c) No, unless you count 'impossible' as the same answer!

19. The number in question must be bigger than 180, but cannot be as big as 180.2, say. So rounding it up to the nearest unit makes it 181.

20. If we are charged a fixed price per unit, fractions of the unit to be counted as a unit, then by rounding up we may calculate the total charge. For example, suppose fencing is only sold per meter and you require 147.2 meters of fencing; you must then purchase 148 meters of fencing.

    If your weight is 147.2 pounds it would be silly to announce your approximate weight as 148 pounds (especially if you're on a diet).

21. 38, $\boxed{3.8 \times 10^1}$
    $\boxed{3.8}$ (note that 3.8 is the same as 3.8×10^0), and 38×10^{-1}
    380, $\boxed{3.8 \times 10^2}$
    .0038, 38×10^{-4}, $\boxed{3.8 \times 10^{-3}}$
    3800, 0.38×10^5, $\boxed{3.8 \times 10^3}$
    .038, $\boxed{3.8 \times 10^{-2}}$

Exercise 4.3, p. 167

1.

School A \ School B	131	132	133	134	135	136	137	138	139	140
111	242	243	244	245	246	247	248	249	250	251
112	243	244	245	246	247	248	249	250	251	252
113	244	245	246	247	248	249	250	251	252	253
114	245	246	247	248	249	250	251	252	253	254
115	246	247	248	249	250	251	252	253	254	255
116	247	248	249	250	251	252	253	254	255	256
117	248	249	250	251	252	253	254	255	256	257
118	249	250	251	252	253	254	255	256	257	258
119	250	251	252	253	254	255	256	257	258	259
120	251	252	253	254	255	256	257	258	259	260

(a) 100

(b) 45

(c) 55

(d) 45

(e) 55

(f) $\dfrac{45}{100} = \dfrac{9}{20}$

(g) $\dfrac{55}{100} = \dfrac{11}{20}$

2. The *answers* are exactly the same, but in the *questions* we must replace 250, 25, 26, by 90, 9, 10, respectively.

3. For the questions we must replace 250, 25, 26 (in Problem 1) by 54, 9, 10, respectively. The answers are

(a) 36

(b) 15

(c) 21

(d) 15

(e) 21

(f) $\dfrac{15}{36} = \dfrac{5}{12}$ (g) $\dfrac{21}{36} = \dfrac{7}{12}$

4. (a) From 0 to 2

(b) $\dfrac{1}{2}$ (by symmetry)

(c) $\dfrac{1}{2}$

(d) $\dfrac{1}{2}$ (This is essentially the same as question (b).)

(e) $\dfrac{1}{2}$

(f) 0, 1 or 2

(g) 1 (The one in the middle!)

(h) (i) $\frac{1}{4}$, (ii) $\frac{1}{2}$, (iii) $\frac{1}{4}$

(j) X and Y now lie between 0 and $\frac{1}{2}$ instead of between 0 and 1, as in (d), so their sum is just as likely to be closer to 0 as it is to be closer to 1.

(k) If one is small and the other large, then Z is certainly 1 because $X + Y$ then lies between $\frac{1}{2}$ and $1\frac{1}{2}$. If X and Y are both large then by symmetry with question (j) Z is equally likely to be 1 or 2.

(l) The probability is $\frac{1}{2}$ that Z is 0 if X and Y are both small and Z cannot be 0 in any other case. The probability that X and Y are both small is $\frac{1}{4}$. Thus the probability that Z is 0 is $\frac{1}{4} \times \frac{1}{2} = \frac{1}{8}$.

(m) $\frac{1}{8}$

(n) $\frac{3}{4}$ $\left(= 1 - \frac{1}{8} - \frac{1}{8}\right)$

5. (a) 108.45 to 108.55

(b) 71.65 to 71.75

(c) 180.10 to 180.30

(d) 180.1, 180.2, 180.3

(e) 180.2, $\frac{3}{4}$, $\frac{1}{8}$ for each of the other two possibilities.

6. (a) 16.370 to 16.380

(b) 12.540 to 12.550

(c) 28.910 to 28.930 (The last zero could be omitted for all six numbers.)

(d) 28.91, 28.92, 28.93

(e) 28.92, $\frac{3}{4}$, $\frac{1}{8}$ for each of the other two possibilities

7. (a) 7.9230 to 7.9240

(b) 5.3800 to 5.3810

(c) 13.3030 to 13.3050 (The last zero could be omitted for all six numbers.)

(d) 13.303, 13.304, 13.305

(e) 13.304, $\frac{3}{4}$, $\frac{1}{8}$ for each of the other two possibilities.

8. (a) The range for the first number is 9.30 to 9.31; the range for the second number is 6.41 to 6.42. Thus the smallest possible difference is $9.30 - 6.42 = 2.88$, and the biggest possible difference is $9.31 - 6.41 = 2.90$. So the solution can therefore be expressed as 2.89 ± 0.01.

(b) The degree of accuracy is the same in each case, but the *centers* of the ranges of the data are shifted, by the *same amount* for each number. Thus the centers of the range for the big number are 9.305, 9.310 and 9.315; and the centers of the range for the smaller number are 6.415, 6.420 and 6.425.

9. Since the length AC is between 11 and 12 meters and the length BC is between 4 and 5 meters the length AB must be between 6 and 8 meters. Thus there are two possible values for the length AB rounded up to the nearest meter, namely 7 and 8. (We exclude 6 because the length AC cannot actually be 11, so that the difference cannot actually be as small as 6.)

Just as in Problem 4 (and in Problems 5, 6, 7) it is not difficult to see that the possible values of the length AB are distributed *symmetrically* about the number 7. That is, the length AB is just as likely to be less than or equal to 7 as greater than or equal to 7. It follows therefore that 7 and 8 are equally likely as the rounded-up values.

10. As in Problem 9, the length AC is between 12 and 13 meters and the length BC is between 5 and 6 meters. Thus there are now two possible values for the length AB, rounded down to the nearest meter, namely 6 and 7 and, just as in Problem 9, these two possibilities are equally likely.

11. (a) 23.62 ± 0.05

    (b) 79.82 ± 0.09 (or ± 0.10)

    (c) 83.10 ± 0.09

    (d) 13.02 ± 0.54

    (e) $5.72 \times 4.13 = 23.62 \pm 0.05 = 2.362 \times 10^1 \pm 0.05$

    $(1.273 \times 10^1) \times 6.27 = 7.982 \times 10^1 \pm 0.09$ (or ± 0.10)

    $(1.0 \times 10^1) \times 8.31 = 8.310 \times 10^1 \pm 0.09$

    $(1.2 \times 10^{-1}) \times (1.0846 \times 10^2) = 1.302 \times 10^1 \pm 0.54$

12. (a) 7519.26 miles ± 0.03 (or ± 0.04).

    (b) 86.1 km ± 0.03

    (c) 2388.82 m ± 0.03

13. 0.09 ± 0.00, $\dfrac{.005 \times 54.6}{(49.87)^2}$

Exercise 4.4, p. 180

1. 138.4. It would be sensible first to group them, say, in 5 cm bands, and then order within each band.

2. (a) 21, 25. The only method would be tedious counting.

    (b) Yes, the average is a fair indicator here since it occurs near the center of the range and the measurements are not too spread out.

3. There are many different answers, among them weight, socio-economic background, shoe-size, and parents' height.

4. (a) 132.6

    (b) Certainly.

    (c) Probably from very early childhood. Of course there is overlap in the range of possible heights between males and females, but we would expect the difference between the average heights of males and females to be significant at all ages, provided that the males and females were taken from the same statistical population— that is, provided that they did not differ significantly with respect to such contributory factors as socio-economic background and ethnic origin.

    To test this theory we would take samples of males and females of the same age and same general circumstances and record their average heights. By doing this for various ages (say, at 5-year intervals from 0 to 25), and for various sets of circumstances, we would build up a picture of the dependence of height difference on sex.

5. (a) Around 136 cm

    (b) Around 141 cm

    (c) It is hard to foretell his height at age $12\frac{1}{2}$ since he has started to grow rapidly. It would be reasonable to guess at least 152 cm.

    (d) The sharp increase may be due to his taking more exercise, eating more food, attaining puberty or other causes (or combinations of causes).

6. (a) 136.3
   (b) 138.3
   (c) 145.2 (or 145.3)
   (d) 149.4 (or 149.5)
   (e) 150.4 (or 150.5)
   (f) 152.4 (or 152.5)
   (g) All these estimates represent intelligent guesses. Probably the least reliable is (f) since we know that height does not continue to increase linearly with age indefinitely.

7. (a) Yes. When the journey starts no gasoline has been consumed.
   (b) Considerably more gas was consumed over the first hundred miles than over the second (and subsequent) hundred miles, so the graph is not linear near the origin.
   (c) Answers will vary. Obvious factors are the condition of the engine; city driving vs. freeway driving; hilly country vs. flat country; much stopping and starting vs. driving straight through.

8. (a) (Answers are approximate, of course.)
   1969: 85,000
   1970: 86,000
   1971: 88,000
   1972: 97,000
   1973: 120,000
   1974: 152,000
   1975: 177,000
   1976: 188,000
   1977: 196,000
   1978: 198,000
   1979: 199,000
   1980: 200,000
   1981: 200,000

   (This assumes that population trend is smooth through the year. There might, in fact, be seasonal fluctuations.)

   (b) Answers will vary, but, typically, new industry will have come into the town, bringing an influx of population. However, availability of space, housing, schools, etc., will put a limitation on the size of the town, so that the growth rate will fall off and the population will stabilize at some figure (here, about 200,000), with the movement in and out of the town effectively balancing. (This scenario ignores birth, marriage and death rates which will probably also be significant.)

9. (a) 137.40; 41.61; 6.45
   (b) 137.66; 60.84; 7.80

10. (a) 7   (b) $\dfrac{35}{6}$

11. (Answers will vary.) It would be useful to compare two given locations for the amount of rainfall or to compare the amount of rainfall at a given location at different times of the year.

    A mere record of the average (without any indication of the spread) would not be useful for telling you how much rainfall to expect at the same place in any given hour of the next day, even if it were reasonable to expect the weather on the following day to resemble fairly closely the weather on the given day.

12. $V(A) = \dfrac{1}{N} \, ((X_1 - A)^2 + \cdots$
    $$+ (X_N - A)^2)$$
    $$= \frac{1}{N} \, (X_1^2 + \cdots + X_N^2$$
    $$- 2ANM + NA^2)$$
    $$= \frac{1}{N} \, (X_1^2 + \cdots + X_N^2)$$
    $$- 2AM + A^2$$

$$= V + M^2 - 2AM + A^2$$

$V(A) - V$ is positive unless $A = M$, when it is 0. Thus the minimum value of $V(A)$ is V, achieved when $A = M$.

Exercise 4.5, p. 183

1. Hint: Think about the areas of the various rectangles in the diagram.

Exercise 5.1, p. 190

1. (a) add
   (b) subtract, larger

2. (a) 12   (g) 13
   (b) 2   (h) −15
   (c) 4   (i) 11
   (d) −27   (j) −21
   (e) 34   (k) 74
   (f) 135   (l) 114

3. Answers will vary, but may include reservoir capacities, temperature, stock and commodities quotations, etc.

Exercise 5.2, p. 197

1. Because they pay more attention to the numbers 1 and 9 than to the words 'billion' and 'million'.

2. 9,000,000 dollars = 9×10^6 dollars, compared to 1,000,000,000 dollars = 1×10^9 dollars. The exponent 9 is much bigger than the exponent 6.

3. (a) 90 years
   (b) 10,000 years

4. Knowing that we're talking about years, and hence writing just the numbers,
   $$90 = 9 \times 10^1 \text{ (or } 9 \times 10),$$
   $$10,000 = 1 \times 10^4 \text{ (or } 10^4).$$

5. The exponents. To see why, note that $3 \times 10^{12} = 3,000,000,000,000$ while $6 \times 10^2 = 600$.

6. A

7. B^2

8. B^7

9. A^7

10. $\dfrac{A^2}{B^5}$ or $A^2 \times B^{-5}$

11. $\dfrac{A^3 \times B^3}{C}$ or $A^3 \times B^3 \times C^{-1}$

12. 1

13. 1

14. $\dfrac{B^2}{A^2}$ or $A^{-2} \times B^2$

15. B^5

16. $\dfrac{B^9}{A^6}$ or $A^{-6} \times B^9$

Exercise 5.3, p. 202

1. The values in the completed table are:

(a)	+4	+2	+8
(b)	+4	−2	−8
(c)	−4	+2	−8
(d)	−4	−2	+8

(b) If the temperature is rising at a rate of 4 degrees per hour then 2 hours ago it was 8 degrees colder.

(c) If the temperature is falling at a rate of 4 degrees per hour then 2 hours from now it will be 8 degrees colder.

(d) If the temperature is falling at a rate of 4 degrees per hour then 2 hours ago it was 8 degrees warmer.

2. The values in the completed table are:

(a)	+3	+4	+12
(b)	+3	−4	−12
(c)	−3	+4	−12
(d)	−3	−4	+12

(a) If a share of IBM stock rises at a rate of 3 dollars each day then four days from now its value will be 12 dollars more than it is now.

(c) If a share of IBM stock goes down at a rate of 3 dollars each day then four days from now its value will be 12 dollars less than it is now.

(d) If a share of IBM stock goes down at a rate of 3 dollars each day then four days ago its value was 12 dollars more than it is now.

3. $\dfrac{A^{10}}{B^{10}}$ or $A^{10} \times B^{-10}$

4. 1

5. A^3

6. A

7. $\dfrac{B^3 \times C}{A^2}$ or $A^{-2} \times B^3 \times C$

8. $\dfrac{B^3 \times C}{A^2}$ or $A^{-2} \times B^3 \times C$

9. A^6

10. A^6

11. $\dfrac{1}{B^8}$ or B^{-8}

12. $\dfrac{1}{B^8}$ or B^{-8}

Exercise 5.4, p. 213

1. 243   **2.** 32   **3.** 6

4. 15   **5.** 6   **6.** 64

7. 1   **8.** 9   **9.** 1

10. 25

11. We know that $(A^{1/2})^2 = A$ and $(C \times D)^2 = C^2 \times D^2$. So if $C = A^{1/2}$, $D = B^{1/2}$, we have
$(A^{1/2} \times B^{1/2})^2 = (A^{1/2})^2 \times (B^{1/2})^2$
$= A \times B$
Thus $A^{1/2} \times B^{1/2} = (A \times B)^{1/2}$. In general $A^r \times B^r = (A \times B)^r$, for any rational number r.

12. $\dfrac{-7}{45}$

13. $\dfrac{9}{50}$

14. $\dfrac{-1}{5}$ or $\dfrac{-11}{55}$

15. $\dfrac{13}{60}$

16. $\dfrac{-3}{13}$ or $\dfrac{-15}{65}$

17. −5432.

18. 6543

19. −7654

20. 8765

21. −9876

22. −8, with a remainder of 2, since $-22 = (-8) \times 3 + 2$.

23. −7, with a remainder of 1, since $22 = (-7) \times (-3) + 1$.

24. −5, with a remainder of 1, since $-29 = (-5) \times 6 + 1$.

25. −4, with a remainder of 5, since $29 = (-4) \times (-6) + 5$.

26. −6, with a remainder of 3, since $-27 = (-6) \times 5 + 3$.

27. −5, with a remainder of 2, since $27 = (-5) \times (-5) + 2$.

28. 1, with a remainder of 1, since
$5 = 1 \times 4 + 1$. (Sorry to insult
you!)

29. 2, with a remainder of 3, since
$-5 = 2 \times (-4) + 3$.

Exercise 6.1, p. 220

1. Let $N = 598283$. Then the answers
and reasons are
   (a) Yes, being $N + 14$
   (b) Yes, being $N - 70$
   (c) No, being $10 \times N + 4$
   (d) No, being $N + 6,000,000$
   (e) Yes, being $N + (10^6 \times N)$

2. (a) 333,334
   (b) 250,001
   (c) 200,001
   (d) 111,112

3. (a) 2^{14}
   (b) 3×7^3
   (c) $2^6 \times 5^6$
   (d) $2 \times 3 \times 5$
   (e) $2^8 \times 3^4$
   (f) 13^2

4. (a) 3^2
   (b) 37^{26}

5. (a) No, $s(1111) = 4$.
   (b) No, $s(11111) = 5$.
   (c) Yes, $s(111111) = 6$.
   (d) Yes, $s(278872278) = 3 \times 17$.
   (e) Yes, $s(123456) = 21$.
   (f) Yes, $s(4176) = 18$ (obviously
   we allow the quotient to be
   a negative integer).

6. (a) No, $s(428387) = 32$, $s(32) = 5$.
   (b) Yes, since $10^{20} - 1$ is a
   string of nines.
   (c) No, since this number is 4
   bigger than the number in
   (b).
   (d) No, since the only prime
   factor of 7^{27} is 7.
   (e) Yes, since $s(123456789) = 45$.

7. X and Y must be the same (if X
and Y were integers, positive or
negative, we could only infer that
$X = \pm Y$).

8. If $X = 3 \times n$ then $X^2 = 9 \times n^2$.
If, on the other hand, X is not
divisible by 3 then 3 would not
occur in the prime factorization
of X and so it would not occur
in the prime factorization of any
power of X. So X^2 could not
even be divisible by 3.

Exercise 6.2, p. 226

1. (a) $2^5 \times 3 \times 7$
   (b) $2^{10} \times 3$
   (c) $3^2 \times 5^5$
   (d) $2^{15} \times 3^2 \times 7$
   (e) $2^{33} \times 3^3 \times 5^3$

2. (a) 2173   (b) 2173
   (c) 2173   (d) $2^5 \times 3$
   (e) $2^{10} \times 3$ (f) $2^{10} \times 3^2$

3. 36 square inches

4. If we want $\gcd(A, B, C)$ we take
the prime factorizations of A, B,
C. If the prime p appears in all
factorizations we take the smallest power in its three appearances
and that is the power of p appearing in the prime factorization of
$\gcd(A, B, C)$.

5. If any number divides A and B it
must divide $A - (B \times Q)$, that is,
R. Conversely if any number
divides B and R it must divide A.
Thus $\gcd(A, B) = \gcd(B, R)$. If
$R = 0$, then B divides A so
$\gcd(A, B) = B$. It is perfectly
consistent to write, for any whole
number B, $\gcd(B, 0) = B$. Repeating
the argument, $\gcd(B, R) =$
$\gcd(R, R_1)$. Therefore $\gcd(A, B) =$
$\gcd(B, R) = \gcd(R, R_1)$. If $A =$
5952, $B = 2952$, then $R = 48$,
$R_1 = 24$, $\gcd(A, B) = 24$.

Exercise 6.3, p. 229

1. (a) 2173
   (b) 4346
   (c) 13038
   (d) $2^{10} \times 3 \times 7$
   (e) $2^{10} \times 3^2 \times 7^2$
   (f) $2^{20} \times 3^2 \times 7^2$

2. (a) $\dfrac{61}{240}$ (b) $\dfrac{13}{336}$

   (c) $\dfrac{53}{13038} = \dfrac{1}{246}$

   (d) $\dfrac{496}{3000} = \dfrac{62}{375}$

3. 560 inches

4. Given any three numbers A, B, C, we find their prime factorizations. If p is a prime factor of any of A, B, C, we take the largest power to which it appears (if it does not appear in a factorization of one of the numbers we say that this power is 0). Then this largest power is the power of p appearing in the prime factorization of lcm(A, B, C).

5. Let p be a prime factor of A or B or both. Let it appear to the power a in A ($a = 0$ if it doesn't appear) and to the power b in B ($b = 0$ if it doesn't appear). Obviously one of a, b is the smaller—call this power s; and one of a, b is the larger—call this power ℓ. Then p^s is the power of p in the factorization of gcd(A, B) and p^ℓ is the power of p in the factorization of lcm(A,B). Thus $p^{s+\ell}$ is the power of p in the factorization of gcd(A, B) $\times$ lcm(A, B) and p^{a+b} is the power of p in the factorization of $A \times B$.

Since, obviously, $s + \ell = a + b$, p occurs to the same power in the prime factorizations of each of these products. This being true for any prime p concerned, the result follows. (Try this yourselves on some special cases, such as the numbers in Exercise 2 of Section 6.2.)

Exercise 6.4, p. 235

1. (a) 7   (b) 8
   (c) 0 (Note that it is enough to observe that the first factor is divisible by 9.)
   (d) 1 (Note that $8 \equiv -1$ and -1 raised to any *even* power is 1.)
   (e) 0 (Note that $7 \equiv -2$ and $(-2)^{12}$ equals 2^{12}.)

2. (a) and (b) may be seen to be false by casting out 9's, (c) and (d) may not. However, (d) may be seen to be false since $512 \times 8172 \times 903$ is obviously divisible by 4, whereas 4001216022 fails the test for divisibility by 4.

3. (a) Certainly $4 \times 7 \equiv 1$. But if we give A any other possible value (0, 1, 2, 3, 4, 5, 6, 8) then $4 \times A \not\equiv 1$.
   (b) A similar argument works.

4. (a) $\dfrac{2}{7} \equiv 8, \dfrac{93}{104} \equiv 6, 8 + 6 \equiv 5$;

   but $\dfrac{12}{13} \equiv \dfrac{12}{4} = 3$

   (b) $\dfrac{5}{19} \equiv 5, \dfrac{8}{23} \equiv 7, \dfrac{15}{11} \equiv 3,$

   $5 \times 7 \times 3 \equiv 6$; but

   $\dfrac{29}{209} \equiv \dfrac{2}{2} = 1.$

Index